D0910230

The Social Gamble

**Determining Acceptable Levels
of Air Quality**

Richard J. Tobin
State University of New York
at Buffalo

Lexington Books
D.C. Heath and Company
Lexington, Massachusetts
Toronto

Library of Congress Cataloging in Publication Data

Tobin, Richard J
  The social gamble.

  Includes bibliographical references and index.
  1. Air—Pollution—United States.   2. Air—Quality
I. Title.
HC110.A4T6   614.7'12   78-19229
ISBN 0-669-02468-6

Copyright © 1979 by D.C. Heath and Company

Published simultaneously in Canada

Printed in the United States of America

International Standard Book Number: 0-669-02468-6

Library of Congress Catalog Card Number: 78-19229

To Dad, Dori, and Ed—
Three of the Best

Contents

List of Figures and Tables

Acknowledgments

Like most other books, this one benefited from the assistance of many people. Countless people on congressional staffs and in several federal agencies were especially helpful in my collecting the data. Similarly the staff in the Documents Room in Lockwood Library at the State University of New York at Buffalo, under the direction of Fred Henrich, provided unsurpassed assistance.

When a draft of the book was completed, several people graciously read and commented on it. J. Clarence Davies, Walter Rosenbaum, Richard Liroff, Beverly Paigen, and Sanford Miller each provided invaluable suggestions that clarified, corrected, improved, and strengthened the presentation. I doubt that I could have received better assistance than these friends provided. The best compliment I can given them is to recommend their considerable services to others. For the typist, Lorraine Belczak, no recommendation is necessary. The excellent quality of her work ensures that her skills will always be in demand.

My wife, Dori, deserves a special note of appreciation. Not only did she type the first drafts, but she also ensured that I would have many quiet hours to write. She sacrificed much of her own time to provide this solitude. Perhaps I can repay her with many games of tennis, walks in the park, and Sunday afternoon rides. She richly deserves them all.

This book was started during a period in which I was receiving financial support from the Research Foundation of the State University of New York. The phrase used as the book's title came from Samuel E. Epstein and Dale Hattis's chapter, "Pollution and Human Health," in W.W. Murdoch, ed., *Environment: Resources, Pollution and Society*.

Introduction

If a government believes that its citizens are exposed to harm, does it have a responsibility to provide suitable remedies? In the United States, the national government's response to this question has been inconsistent. It has banned the production of some harmful products such as cyclamates while it has encouraged the use of other harmful products such as tobacco.

Different responses to such hazards raise several questions. How can a government encourage the use of one item when it bans another product that allegedly is of lesser harm? What criteria should a government use when it makes these decisions? How does a government determine how harmful some products are? How much and what kinds of scientific data should a government have before it initiates regulatory action? If regulation is in order, what values should affect the pace and nature of remedial action?

Any government that attempts to protect its citizens from environmental pollutants must resolve these questions. In terms of air quality, as an illustration, the Congress has passed many laws with the intent of protecting the public's health and of guaranteeing a high quality of life. When Richard Nixon signed the Clean Air Amendments of 1970, for example, he declared the law to be the most important air quality legislation in the nation's history, and he said that it would "put us far down the road toward a goal . . . of clean air, clean water and open spaces for the future generations of America."[1]

In this instance, the president's statement contained an implicit assumption that a close relation exists between the intent of a law and its results. The public often is accused of just such a belief. As Marver Bernstein has noted, a tendency exists to regard implementation as automatically following legislation: "If statutory authority is available, the expectation is that it will be exercised."[2] In contrast to this position, much recent research indicates that great discrepancies often exist between legislative intent and actual outcomes. Jeffrey Pressman and Aaron Wildavsky's study of the Economic Development Administration in Oakland and Martha Derthick's short critique of governmental efforts to create new communities offer two of the best examples of this divergence.[3] Their research emphasizes that successful implementation does not necessarily follow a law's passage or a policy's formulation. This finding raises an important concern. Many politicians believe that their responsiveness to the nation's problems can best be measured by counting the number of new programs initiated during their tenure in office. Despite the pervasiveness of this belief, the ultimate test of a government's responsiveness lies not in the passage of laws but in their successful implementation.[4]

Perhaps the most important aspect of any implementation process is the meaning assigned to policy goals. Without a precise understanding of what constitutes fair housing, equal opportunity, or environmental quality, legislative mandates and presidential proclamations calling for acceptable levels of these values are meaningless. Statements of preference must be translated into guidelines for action, and *The Social Gamble* explores this process in terms of the definition of clean air. The book is based on the premise that the achievement of clean air is less dependent on how laws or politicians extol its virtues than on what administrators do about it. Accordingly the book examines and analyzes how the federal government, once it had decided to abate air pollution, has determined the risks associated with this pollution and translated a vaguely stated preference for clean air into a precise definition. The first task clearly belongs in the realm of science, but because *clean air* has no scientific meaning, the second task falls to public officials, who must decide how clean is clean enough. Such choices usually are made on the basis of expected benefits or on social and economic considerations. Regardless of the factors used, however, the final decision will reflect a normative judgment.

To a purist, such a judgment means the absence of all pollutants. To achieve this level of cleanliness would require an end to all industrial activity, as well as the end of current life-styles. More realistically, a definition of clean air provides for some pollution, but the question is how much. The answer is highly significant. How *clean air* is defined is crucial to the public, to polluters, and to the potential impact of any control program. The definition will determine the industries and life-styles affected, the design of abatement programs, the kinds of control technologies employed, and the use and distribution of such scarce energy supplies as natural gas and low-sulfur oil. Similarly the definition plays an important role in determining the quality of life, the extent of public support for abatement programs, and the public's perceptions of governmental responsiveness to an important national program. Depending on the choices made, people will be subjected to continuous levels of harmful pollutants or protected from an invidious list of respiratory and cardiovascular ailments. Finally a definition of clean air establishes a goal against which a clean air program's progress and success can be measured and evaluated.

The need for a specific goal means that some choices must be made and that some groups must be charged with the responsibility for defining clean air (or, alternatively, for defining the existence of air pollution). From some perspectives, legislators should accept this responsibility and pass laws accompanied by clear standards of implementation.[5] If legislators would do this, administrators could compare scientific judgments on the harmful effects of air pollutants with precisely drawn legislative requirements. The result would be clear-cut guidelines, congressional assent, and general public acceptance.[6] Equally important, administrators would not be required to defend what others believe to be capricious value judgments. Despite these seeming advantages for administrators, legislators often pass laws that leave the precise definition of goals and standards to administrative agencies. Such vagueness is often deliberate and intended to facilitate compromise among legislators, but it is not enough for the Congress to instruct one agency to protect the public from monopolistic practices or to require another agency to protect consumers from unsafe products. Reluctance or inability to provide specific guidelines ordinarily means that administrators must mediate the political conflicts that congressmen have avoided, but this buckpassing is not without costs. Bernstein states the problem well: "In the absence of legislative declaration of goals which regulatory policy should follow, a regulatory agency will function without benefit of political compass and without adequate intelligence or supply lines."[7]

In the case of the nation's clean air laws, congressional action has paralleled Bernstein's observation. The Clean Air Act of 1963, the nation's first permanent air pollution legislation, was designed to protect the nation's "air resources so as to promote the public health and welfare and the productive capacity of its population."[8] Subsequent legislation has repeated this general statement of purpose, and administrative agencies have been required to define acceptable levels of air quality. Yet they cannot define a single level of acceptability for all air pollutants. There are far too many detrimental pollutants from a very large number of sources. To examine how administrators define acceptability for all these pollutants would be both confusing and of little value. Alternatively focus on a single, important category of pollutants seems more appropriate. This study adopts just such an approach and looks at what are probably the most damaging and most controversial of all pollutants: sulfur oxides (primarily sulfur dioxide).

Notes

All publications of the federal government and of the National Academy of Sciences were published in Washington, D.C., unless otherwise noted. The following abbreviations are used in the notes:

CEQ   Council on Environmental Quality

EPA   Environmental Protection Agency

FEA   Federal Energy Administration

FPC   Federal Power Commission

HEW   Department of Health, Education and Welfare

NAS   National Academy of Sciences

NAPCA   National Air Pollution Control Administration

OMB   Office of Management and Budget

PHS   Public Health Service

1. *Public Papers of the Presidents of the United States: Richard M. Nixon—1970* (1971), pp. 1166, 1168.

2. Marver H. Bernstein, *Regulating Business by Independent Commission* (Princeton: Princeton University Press, 1955), p. 82.

3. Jeffrey L. Pressman and Aaron Wildavsky, *Implementation* (Berkeley: University of California Press, 1973), and Martha Derthick, *New Towns In-Town* (Washington, D.C.: The Urban Institute, 1972).

4. Bernard J Frieden and Marshall Kaplan, *The Politics of Neglect* (Cambridge, Mass.: MIT Press, 1975), p. 3.

5. Theodore Lowi, *The End of Liberalism* (New York: W.W. Norton, 1969), p. 298.

6. Peter Barton Hutt develops this theme in "A Regulator's Viewpoint," in NAS, *How Safe Is Safe?* (1974), pp. 116-131.

7. Bernstein, *Regulating Business,* p. 286.

8. PL 88-206, sec. 101(b)(1).

1

Sulfur Dioxide, Politics, and Science

Coal and oil are two of the country's most important energy resources; they are also the major source of one of the nation's most harmful pollutants. When coal and oil are burned, nearly 95 percent of the sulfur in coal and all of the sulfur in oil are released into the atmosphere. Some of the sulfur is released in the form of sulfuric acid, but the great bulk of emissions take the form of sulfur dioxide (SO_2), an acrid, corrosive, and poisonous gas. Once in the atmosphere, SO_2 is gradually transformed into sulfates, although the exact chemical processes are not well understood.

Emissions of SO_2 rose rapidly in the 1960s, from slightly over 23 million tons in 1960 to nearly 34 million tons in 1970, as a result of large increases in the use of coal and oil. A projection made in the early 1970s indicated that if left uncontrolled, emissions of the pollutant would total over 60 million tons in 1980 and nearly 126 million tons in 2000.[1] The increase would be more than 540 percent from 1960 to 2000. Abatement programs introduced in the early 1970s stemmed the increase in many urban areas, but the total loadings of SO_2 continue to rise in many parts of the United States.

The emissions of other pollutants exceed the total emissions of sulfur oxides, but no other pollutant has as much effect on the nation's health.[2] The oxides of sulfur have been implicated as a leading contributor to illness and death in virtually every major air pollution episode of the last fifty years. A report released in 1977 concluded that over twenty thousand people die prematurely every year because of air pollution caused by the burning of coal and oil at electric power plants.[3] This estimate is probably conservative since the figures include only states east of the Mississippi River. According to the study, which was conducted for the National Academy of Sciences, SO_2 and sulfate particles exacerbate existing respiratory problems, bring about lung and heart failures, and shorten some people's lives by as many as fifteen years.

At levels measured in parts per billion, SO_2 has been linked to such lung disorders as chronic bronchitis and pulmonary emphysema. Certain segments of the population are especially sensitive to the pollutant's adverse effects. Extended exposure to low levels of sulfur oxides has been linked to increased rates of cardiovascular morbidity in the aged, and exposure to higher concentrations has been linked to schoolchildren's complaints about coughs, mucous secretion, and mucous-membrane secretion. Sulfur oxides have also been associated with decreased respiratory functions in healthy adults.

1

Sulfur oxides damage not only the public's health but its welfare as well. SO_2 can prolong fogs, reduce visibility, and damage paint, clothing, vegetation, and many building materials. At least one scientist believes that levels of SO_2 are almost high enough in some parts of the United States to "cause precipitous effects on agricultural production."[4] One by-product of SO_2, acid rainfall, causes great harm to plants and animals. It has been cited as a cause of serious fish kills, the increased acidity of many lakes and rivers, and undesirable effects on predatory-prey relationships.

The damages attributed to the pollutant can be enormous. The President's Council on Environmental Quality (CEQ) once estimated that SO_2 annually causes $3.3 billion damage to the public's health and another $5 billion damage to property, materials, and vegetation.[5] A second estimate placed the health damages attributed to sulfur oxides at $5.4 billion in 1977 (in 1970 dollars) and the damages to property, materials, and vegetation in 1977 at more than $8 billion (also in 1970 dollars).[6] One reason why sulfur oxides have such a great impact is that their major sources tend to be located in or near metropolitan areas. Furthermore most emissions occur in the most heavily populated section of the United States, the northeast quadrant.

The detrimental and pervasive effects of SO_2 have led the nation's policy makers to focus much attention and many resources on its control and substantial abatement. A review of expenditures since 1963 reveals that the federal government has spent more money researching the health effects of SO_2 and trying to develop control technology for it than it has for any other pollutant. Despite policy makers' admirable motives and goals, efforts to reduce emissions of the pollutant are among the most conflict-ridden and politically sensitive of all environmental programs. The pollutant's major sources are steel mills, copper smelters, petroleum refineries, and electric power plants. Sulfur oxides are inextricably tied, therefore, to the nation's economic strength, to its industrial viability, and, more recently, to its dependence on foreign supplies of energy and its ability to implement an effective energy policy.

These considerations and the simultaneous desire to reduce levels of SO_2 have frequently led to tension among environmentalists and industrialists. A stagnant American economy in the 1970s aggravated this strain and brought about a reconsideration of the nation's priorities. The Environmental Protection Agency (EPA), which has responsibility for implementing the SO_2 program, has been besieged by criticisms of its operations and implementation of the program. The complaints from the coal and electric utility industries tend to be especially strong.

The National Coal Association has claimed that air quality legislation is unresponsive to the nation's needs because it discourages the use of the country's largest energy resource. Coal accounts for all but 10 percent of America's reserves of fossil fuels, yet its relative importance has declined precipitously since 1950, when it produced almost 40 percent of the nation's energy needs; by

1976, this figure had been cut in half. The decline has not been felt equally among all sectors. As the industrial, transportation, and household-commercial sectors were reducing their consumption of coal, power companies were increasing theirs. The utility industry burned less than 20 percent of all coal mined in the United States in 1950, but this figure had risen to more than 75 percent by 1977.[7] This change creates a situation in which the coal industry's well-being is highly dependent on electric utilities' choice of fuels. Consequently the coal industry is especially sensitive to any environmental regulations affecting this choice.

Although electric utilities have brought about a large absolute increase in their use of coal, the relative contribution of coal to total production of electrical energy began to decline after the mid-1960s. Regulations to abate SO_2 were responsible for much of this decline because they increased substantially the costs associated with coal usage. To comply with these regulations, many companies found burning natural gas or low-sulfur oil easier and less costly. The use of these two fuels is the most important reason for reduced levels of SO_2 in most urban areas. Thus maintenance of existing levels of air quality or further improvements are heavily dependent on the continued availability of natural gas and low-sulfur oil. Paradoxically as these two fuels have been used to bring about lower levels of SO_2, they have become more difficult to secure. Today shortages of natural gas and high prices for oil have created a renewed interest in coal. As part of President Jimmy Carter's national energy plan, consumption of coal is supposed to double between 1976 and 1985 to 1.2 billion tons per year. As one analysis has suggested, however, "The most severe current environmental challenge to [increased] coal use relates to the control of sulfur oxides."[8] The National Academy of Sciences' study referred to earlier puts this statement in perspective. Scientists who took part in the study concluded that a large-scale increase in the use of coal could lead to as many as thirty-five thousand premature deaths per year by the year 2010.[9] These deaths are projected to occur even if power plants install control equipment to cut emissions of SO_2 by 80 percent. In short, the effects of SO_2 become more important as a result of the energy situation and the concurrent plan to burn more coal.

Electric utilities are expected to use most of the increased coal production, but the utility industry has argued that the implementation of the abatement program for SO_2 is an "illogical and a terrible waste of money for the American people."[10] Many of the electric utilities' complaints stem from the enormous costs associated with reducing air pollution. For the entire industry, it has been estimated that $20 billion will have to be invested between 1975 and 1985 to abate air pollution. Most of this money will be spent trying to control sulfur oxides. The total expenditures for environmental protection represent an amount larger than that required for any other industry.[11] It has also been estimated that utilities' expenditures for pollution control will increase the cost of electricity far more than will similar expenditures affect the costs of products from thirty other major industries. In one study prepared for the EPA and the CEQ, utility prices were projected to increase more than 27 percent from 1970 to 1983 as a result of the costs of controlling air pollution.[12] This change compares with much smaller increases for the industries producing nonferrous metals (8.1 percent), paper (6.2), automobiles (5.5), chemicals (4.8), and iron and steel (4.1). The average increase for thirty-one selected industries is expected to be only 2.6 percent. If these figures seem modest, the expected increase for electric utilities is not. Moreover the projected costs probably underestimate the actual costs because the figures were compiled for two government offices with a vested interest in minimizing the economic costs of environmental protection.

For the electric utility industry, pollution control expenditures and higher prices will not be distributed evenly among all companies or all consumers. Instead the costs will fall most heavily in areas where utilities rely on coal as their main sources of energy. From the perspective of these utilities, such expenditures are especially irritating. Many utility executives believe that the problems of air pollution are overemphasized and that the industry already is doing enough to protect the nation's air quality. Even if electric utilities were enthusiastic about pollution control, it is unlikely that many of them have enough readily available capital to comply with federal standards. Among all industries, electric utilities are especially capital intensive. They must borrow large amounts of money but find it difficult to do so during periods of depressed stock prices and high interest rates. Electric utilities are also beset by a great deal of uncertainty. Once a company has estimated the need for a new power plant, it can take up to ten years before the plant is in operation. During this period, changes in life-styles, fuel availability, and government policies can alter demands for electricity. And once a plant is completed, for example, it may be found unnecessary or inadequate. All of these factors help to explain why many power companies are apprehensive about any government policies that increase costs and uncertainty.

In addition to industrial criticisms, efforts to abate SO_2 have caused millions of Americans to evaluate their willingness to pay for improved levels of environmental quality, their dependence on and use of foreign oil supplies, which are frequently low in sulfur content, the desirability of strip-mining vast coal reserves in the western United States, and the continuance of an energy-intensive economy and life-style. To a major extent, these issues arise because policy makers have decided to decrease the amount of SO_2 in the nation's air to what they consider safe or acceptable levels.

Assessment of Risk and Determination of Acceptable Risk

When an administrative agency attempts to define clean air, the process is similar to that used in establishing safe levels for other products or substances.

The process involves science and politics in three related steps. First, scientists generate data about the effects of a pollutant. Second, scientists within the government review the available data and assess the relative risks of exposure; they calculate how much of the population is exposed to the pollutant and what part of the population will be affected by it. After these two initial tasks have been completed, policy makers decide what will be a safe level, and this is a "judgment of the acceptability of risk."[13] Since *acceptability* has no scientific meaning, judgments of this sort can be made only on the basis of habit, perceived benefits, or social, economic, political, or ethical considerations among others. Similarly the process of judging safety or acceptability is a normative matter of personal or governmental preference for which scientists are no better qualified than any other group. Consequently while scientists can indicate relative probabilities of harm, their professional expertise does not give them any advantage in deciding the level of risk that a society should tolerate.

Although the determination of acceptable levels of risk is a political decision, this does not mean that policy makers will eschew scientific data about the probabilities of risk. On the contrary, without such data policy makers would be hard pressed to justify their decisions. Scientists have the greatest familiarity with their research findings, are better equipped to understand the implications of the findings, and are aware of the data's scientific limits. It would seem essential, therefore, that these attributes be incorporated into the policy-making process. Indeed many policy decisions cannot be made without the information that only scientists or professional administrators can provide. On this point, some people have suggested that serious problems can follow a politician's neglect of professional opinion. "From the point of self-preservation alone," says Francis Rourke, "politicians are obliged to lean heavily upon the advice of bureaucratic experts in making policy decisions."[14] In short a public official who ignores scientific advice about potential hazards does so at his own peril.

The need to rely on scientific advice has an important corollary: if political decisions about acceptability are to be effective, they must be based on honest and competent advice. According to Rourke, sufficient "candor can only be secured if professionals are protected from reprisal for policy findings or recommendations that may be offensive to politically potent groups."[15] Strong executive or legislative support for an agency's activities will often deter undue pressure from such groups. In the absence of protection from external pressures, however, bureaucratic advice may hinge less on its accuracy than on its acceptability to affected groups.

However the assessment of risk proceeds, the result is a statement that links different levels of harm or safety to different levels of exposure. In the case of air pollutants, the end result is a set of air quality criteria, which

> are an expression of the scientific knowledge of the relationship between various concentrations of pollutants in the air and their adverse

effects on man, animals, vegetation, materials, visibility, and so on. . . .
Air quality criteria are descriptive—that is, they describe the effects that
can be expected to occur whenever and wherever the ambient air level
of a pollutant reaches or exceeds a specific figure for a specific time
period.[16]

When they produce an overall assessment of risk, scientists have fulfilled their
scientific responsibilities. Further responsibility rests with public officials; it is
up to them to determine levels of acceptable risk.

Deciding upon an acceptable level of risk can lead to a wide diversity of
views. Rene Dubos has been quoted as saying that inordinate concern for safety
can stifle economic growth: "To demand a certified verdict of safety before
accepting a new technological innovation would clearly result in paralysis of
progress."[17] Dubos' opinion typically is applied in the United States. Most new
products are assumed to be beneficial and are marketed until substantial evidence exists that such products cause harm or hazard. Should there be a possibility of harm, the burden of proof to show otherwise is usually not on the
manufacturer but on the user (or a relevant government agency).[18] To do otherwise, claims the National Industrial Pollution Control Council, "could be likened
to a requirement of proof of innocence by the accused rather than proof of guilt
by the accuser."[19]

Garrett Hardin assumes just the opposite view. When a new product is introduced to the market, Hardin says that the likelihood of harm is at least as probable as the likelihood of no harm. If the probability of success is no better than
half, Hardin says that a new product should be judged guilty until exhaustive
tests have shown otherwise.[20]

Both Dubos and Hardin express opinions on how the risks of new products
ought to be handled, but neither gives any direction on how (or even if) products already in use should be evaluated to determine their potential for harm.
Should an item with a long history of use be exempted from scrutiny on the
basis of its seniority and popular acceptance, or should all producers be required
to show that their products are harmless, including those with no commercial
value? Similarly because all substances can be harmful if taken in large enough
quantities, are manufacturers liable only for damages incurred in normal use?
Finally if a case can be made that manufacturers must evaluate the potential
harm of in-use products, what kinds of risks must a manufacturer consider (for
example, should the manufacturer be required to test for genetic defects if his
product is used to relieve earaches), and how often must these risks be considered (for example, if a product was judged to be free of harm in 1940, does
that justify its continued use forty years later)? It is well known that many
products and substances now on the market are there on the basis of earlier
tests, which are considered inadequate by modern standards. Further many new

tests exist today, but they have not been applied to most in-use substances or products.

For these in-use products, the government's usual response is to allow them to remain on the market. However one views this strategy, it is probably satisfactory to a large number of people on economic grounds, if for no other reason. The cost of measuring risk for all in-use products would be staggering, and the need for testing would be fought vigorously by producers who could claim that testing would make no sense since many years of prior use had not led to any harm. In addition, because new scientific tests are invented regularly, it is infeasible "every year to go back and retest, using the latest methodology, all the components of food and drugs [or any other products] that previously have been placed on the market."[21] More important, the best that scientific tests can do is to demonstrate the existence of harm and not its absolute absence.[22]

These problems further illustrate that although scientists can be asked to indicate the probability of risk, nonscientific values are used to judge the acceptability of this risk. Alvin Weinberg indicates that such situations are transscientific: scientists are asked questions that science cannot answer. Under these circumstances, Weinberg believes that the translation of scientific findings into safety standards should be handled in the same way that other public policies are decided—through an adversary process.[23] An adversary process implies a balancing of objectives, which can be enormously difficult.

Advocates of the status quo are easily found, and attempts to reduce a substance's presence will confront well-established efforts to maintain existing commercial or industrial patterns. Opponents of change do not find it difficult to estimate potential financial losses if their products are banned, limited, or regulated, but scientists cannot place a monetary value on human health or well-being, especially when the precise relation between a product and levels of harm is only poorly known. Thus when risks and benefits are not comparable, application of risk-benefit analysis is questionable.

In view of this uncertainty, prudence might dictate a cautious approach that errs on the side of safety instead of on the side of economic or commercial benefits. Such a course has certain appeal, but the consequences of a wrong decision can be both expensive and embarrassing. For example, in the late 1960s, the federal government became concerned about the rapid eutrophication of many of the nation's lakes due to heavy use of phosphates in detergents. The government pressured manufacturers to shift from phosphates to nitrilotriacetic acid (NTA) and, by 1969, the government had demanded its use, even though it knew little about NTA's potential side effects. In response to this pressure, private industry spent millions of dollars in developing NTA and in gearing up for its widespread production. By 1970, however, the government had subjected the detergent industry's "great white hope" to careful examination and had found that birth defects might occur if NTA entered a water supply. Within a few weeks of this finding, the government requested that further use of the acid be terminated.

The losses associated with NTA affected only one industry. In the area of air pollution, however, a wrong decision would have far greater consequences. Too lenient standards could jeopardize the health, lives, and productive capacity of millions of Americans. In contrast, if a definition of clean air is unnecessarily stringent, industry and consumers will waste billions of dollars on economically nonproductive and unnecessary pollution control equipment. The fact that a direct relationship does not exist between reductions in pollution and abatement costs magnifies this situation. The first increments of pollution are the cheapest and easiest to abate, but as a polluter seeks to reduce further increments, his costs and problems spiral. As figure 1-1 illustrates, a polluter might find that 70 percent of his emissions can be reduced at a cost of $10 million. If pollution standards are tightened to require another 10 percent reduction (to 80 percent), the polluter will face additional expenses of $5 million. In the end, he has achieved only a marginal reduction in additional pollution at a major expense.

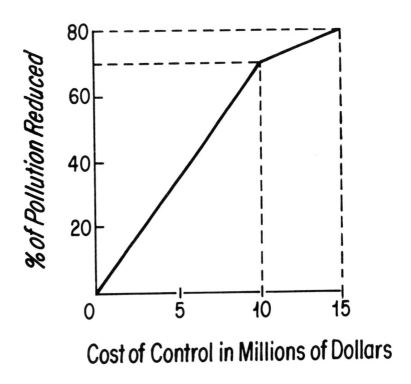

Figure 1-1. Relation between Cost of Controls and Pollution Abated

Such circumstances have led many people to conclude that environmental standards, particularly those based on tenuous data, waste the country's resources and reduce its potential for growth. In short, decisions about acceptable levels of risk are clearly authoritative allocations of values because every decision about safety rewards some people but penalizes others.

Guides to Acceptability

Although decisions about acceptability must be political, this does not mean that guides to acceptability are absent. In *Of Acceptable Risk*, William Lowrance has provided a useful catalog of some current ones.[24] These include reasonableness, custom of usage, prevailing professional practice, and degree of benefit or necessity. These and other similar guides share a common weakness because they all require subjective decision making. For example, the benefit-necessity guideline assumes that certain items or products should be allowed to remain on the market as long as their benefits exceed their costs. Just as beauty exists in the eyes of the beholder, however, an item's value depends on who is using it and for what purposes.

In addition to the subjective guidelines, Lowrance indicates that several empirical criteria are available for application. Two—exposure relative to natural background and occupational exposure precedent—use background levels of noise, pollution, and similar irritants as a basis for comparison. With the first guideline, an assumption is made that if exposure to a certain agent is kept well below the natural background level for that agent, "we will be assured that the additional consequences will neither differ in kind from those which we have experienced throughout human history nor exceed them in quantity."[25] With the second guideline, occupational experience is used to determine levels of maximal exposure without harm or adverse effect. A generous safety factor is then applied to this occupational exposure in order to establish a limit for the general public.

Both of these guidelines have weaknesses. First, policy makers assume that background levels are inherently acceptable, which may not be the case. The incidence of skin cancer due to exposure to sunlight provides an example of a situation in which natural levels can be harmful. Second, the extrapolation of acceptable occupational exposures to the public can also be questioned. Safe exposure for healthy male factory workers over an eight-hour, five-day workweek cannot be generalized to a population composed of all ages, two sexes, and healthy and unhealthy people continually living in the same environment.

A final procedure can ignore all scientific data and rational discussion yet can be the most appealing and is certainly the most democratic. This approach simply asks people to vote their preferences for acceptable levels of risk through public referenda. In order to be effective, however, such voting would have to be on clear-cut choices (should the community fluoridate its water supplies?) and not on a large number of discrete choices (what should be acceptable levels of air pollution? for specific pollutants? for what time periods or meteorologic conditions?). It is also unlikely that such voting would be useful for very large jurisdictions, such as regions or states that have diverse environments. People are more likely to tolerate polluted air in a city's industrial core than they are in a national park, but the park should not be inflicted with the pollution merely because the city has more voters.

Though many guidelines for setting safety standards exist, none are completely satisfactory. A conscientious policy maker faced with a need to set a safety standard or to define acceptable levels of air quality would be hard pressed to justify a single guideline to the exclusion of others. Both the choice of criteria and the criteria themselves require that social, economic, and political preferences play a leading role.

Whatever the means used to determine acceptable levels of risk for air pollutants, the end product is a definition of clean air, and this is expressed as an air quality standard, a numerical statement of goals or desired levels of air quality that prescribes "pollutant levels that cannot legally be exceeded during a specific time period in a specific geographic area."[26] Usually, but not necessarily, a relation exists between air quality criteria (the assessment of risk) and air quality standards (a statement of acceptable risk). For example, if scientific research indicates that SO_2 causes ill effects at levels of x parts per million over a period of twenty-four hours, then policy makers could set an air quality standard to limit the amount of SO_2 in the ambient air to what is considered an acceptable level. Without a criteria document or some other kind of summary, public officials would have no way to determine when a pollutant affects the public's health, and in this manner air quality criteria provide the credibility and scientific basis for air quality standards. It is important to remember, however, that the setting of air quality standards is a political, and not a scientific, responsibility. This means that there may be some cases in which air quality standards bear no relation to the criteria. Policy makers can ignore assessments of risk for such reasons as cost, political opposition, infeasibility of attainment, or even lack of desire.

Applying a Definition of Clean Air

After scientists have compiled a criteria document and after policy makers have selected an air quality standard, which defines an acceptable level of risk (a definition of clean air), the next step is the application of this definition to an abatement program. The definition of clean air is applied in at least two ways. First, air quality standards provide goals for an abatement program. Accordingly when a standard is selected, officials know how much of the existing pollution must be reduced before clean air is achieved. A definition, there-fore, provides a gauge of progress and a measure of successful (or unsuccess-ful) policy implementation.

Second, a definition of clean air allows policy makers to establish an abate-ment program. Air quality standards create goals, but the standards do not indicate how to achieve these goals. Nonetheless if officials know the amount of pollution emitted into the air and the level of air quality that results from these emissions, they can estimate the quantity of emissions that must be reduced to achieve the air quality standard. To achieve such a reduction, offi-cials set emission standards that specify the amount and kind of pollutants that can be emitted from a source of pollution. Emission standards, which are usually set so that air quality standards can be met, indicate to polluters that they cannot legally discharge more than y parts of a pollutant into the air or that they cannot use fuel containing more than x percent of sulfur or z pounds of sulfur per million British thermal units (Btu).[27] Emission standards will vary from one place to the next because of local differences in the amount of pollu-tion, the kinds and numbers of polluters, and the degree to which officials wish to incorporate a concern for aesthetics or other values.

Even without local differences, however, the determination of appropriate emission standards can be risky and contentious. In setting these standards, officials must make at least the following assumptions:

1. That an accurate inventory of pollutants exists, which must include the amount and source of the pollutants. An adequate inventory requires complete cooperation from all polluters. In the case of sulfur oxides, each user of oil and coal must provide comprehensive information on the amount of fuel used and its sulfur content.
2. That the amounts and kinds of pollutants in the air can be measured accu-rately. This is frequently not the case.
3. That a reduction in emissions will cause a proportionate improvement in air quality. At best, this is a questionable assumption since such considerations as meteorological and topographical elements affect the dispersion of pollutants.[28]

Should there be difficulties with any of these assumptions, and there often are, then the utility and credibility of any emission standard will be in doubt. Consequently the process of translating air quality standards into emission standards can be a highly politicized task.

Emission standards are important not only as a means to reduce pollution but also as a basis for all enforcement programs. As such, emission standards regulate and prescribe polluters' actions. If the analyses of Theodore Lowi and others are correct, then such regulatory policies as emission standards constitute an arena of power. Lowi has written that different arenas tend to develop their "own characteristic political structure, political process, elites and group relations."[29] In other words, he believes that knowledge of a public policy's impact allows one to make predictions about the nature of the political situation surrounding that policy. These forecasts ought to be useful in explaining and understanding the politics surrounding the definition and application of clean air.

According to Lowi's model, a general pattern of regulation develops in which Congress establishes broad goals and then requires an administrative agency to provide the specific details and guidelines. Although regulatory policies tend to be stated in general terms, their impact is to affect directly the profits and operations of those subject to regulation. Such regulation rewards some groups at the expense of others, so the competition for these rewards is intense, or so it is argued. In addition, the influence of affected interest groups is high, and coalitions of these grouips try either to avoid regulation altogether or, "where regulation is unavoidable, they . . . press for regulation that is either favorable or at least not too bothersome to their interests."[30]

In the regulatory arena, the potential for conflict is exceptionally high because regulation can adversely affect interest groups' well-being. This conflict is as likely to be between interest groups and administrators as it is between congressmen and administrators. On the one hand, "regulation always brings the possibility of protest and reprisal." On the other hand, conflict is likely when "congressional actors who are seeking exceptions to regulations for favored constituents or clients meet with resistance from bureaucrats who oppose these exceptions."[31]

It is also likely that such conflict will involve the entire Congress and many key officials in the executive branch. Should affected parties lose a regulatory battle in a congressional committee or in an administrative agency, an appeal to the entire Congress or to the president is a distinct possibility. Finally changing social and economic conditions and the election of new presidents and congressmen mean that regulatory issues normally are assured a permanent place on a government's agenda. As a result, regulatory laws or guidelines are rarely final; they are subject to continuous renegotiation and readjustment.

The political characteristics of regulatory policies are likely to be especially evident in the area of environmental protection. Unlike traditional forms of regulation that establish limited controls over rates and markets, environmental protection can be seen as a form of social regulation in which regulators attempt to ensure that economic interests consider the health, safety, or environmental costs of their activities. To accomplish this goal, social regulators become intimately involved with now goods and services are designed and produced. As an illustration, environmental regulations on air quality often specify in great detail what kinds of fuels industries can use, how and when these fuels can be burned, and the kinds of control equipment that polluters must install. Many businessmen have long considered such choices to be the prerogative of private enterprise, but under social regulation these choices often are subject to bureaucratic consent. Social regulators thus become important determinants of product cost and of the choices available to consumers.[32] In short, regulated interests have fewer degrees of freedom under social regulation than they do under traditional modes of economic regulation.

Economic regulation has long been the subject of scholarly discussion, but just the opposite is true for social regulation. As an example, Lowi's seminal article on arenas of power was written several years before the creation of such social regulators as the EPA, the Consumer Product Safety Commission, and the Occupational Safety and Health Administration, and therefore made no distinction between the two forms of regulation. Only recently have scholars begun to focus on the implications of social regulation. Thus relatively little is known about the extent to which social regulation shares the characteristics, political processes, and impacts of its economic colleague. The phenomenal growth in the number of social regulations in the 1970s and the likelihood of their continued increase in the future strongly suggests the need to remedy this deficiency.

A quick glance at the pervasiveness of and concern for environmental regulations further demonstrates the utility of such an examination. Few industries, institutions, or individuals find themselves free of the restrictions that environmental agencies impose. Indeed the reach of environmental programs affects and shapes the entire economy, as well as the operations of government agencies at all levels. Few other regulatory programs can claim this distinction. A few years ago one congressman humorously estimated that it would require seven hundred thousand people to implement existing environmental legislation at the national level. In a more serious mood, the congressman added that the EPA "is mandated to do more in more ways and regulate more than any one agency in our history has been called upon to do."[33]

The fact that environmental regulations can restrict the activities of so many interests is one source of controversy, but another is the fact that the benefits of environmental protection are rarely certain, often tentative, and usually difficult to quantify. For many people, this reduces the appeal of environmental protection. Moreover in order to achieve the benefits of this protection, changes in habits and life-styles must come about now, although the rewards of change may not be evident for several generations. In contrast to the elusive benefits, the economic costs of environmental regulation tend to be large, immediate, and easily quantified. These traits provide an advantage to critics who claim that the costs of environmental protection far exceed its benefits.

These considerations help to explain why environmental programs tend to stimulate both meticulous scrutiny and intense opposition. A brief review of congressional interest in the EPA offers several revealing illustrations of this point. As the federal government's largest regulatory agency, the EPA is subject to the direct jurisdiction of at least five committees in the Senate and seven in the House of Representatives. More than twenty other congressional committees claim some partial jurisdiction. The members of these committees have not been at all reluctant to focus their attention on the EPA. Between 1973 and mid-1976, for example, officials from the agency testified at 208 separate congressional hearings.[34] Over the same period, no other regulatory agency appeared as frequently. Officials of the second most studied agency, the Federal Energy Administration, appeared only 120 times, even though energy policy was at the center of the nation's attention throughout the period.

Legislative hearings are not the only way in which congressmen express their opinions about environmental regulations. Since its creation in 1970, the EPA has faced frequent congressional intervention in its decision making, attempts to hamstring its operations, efforts to transfer its functions to other agencies, and even legislation designed to abolish it altogether. Although Congress's two appropriations committees ordinarily do not become involved in substantive legislation, this has not been true in regard to the EPA. In successive years, the House Committee on Appropriations recommended that the EPA be required to bear the costs of all environmental impact statements written in conjunction with the National Environmental Policy Act (1972); allocated $5 million so that the National Academy of Sciences could conduct a "complete and thorough review, analysis, and evaluation of the [EPA], its programs, its accomplishments, and its failures" (1973); and proposed that the agency not spend any money on programs that would reduce the availability or increase the cost of food or electricity (1974).[35]

In sum, environmental regulations provoke controversy and extensive anxiety on the part of polluters and public officials, and environmental agencies must pacify and diffuse these reactions. One of the best ways to achieve this goal is through unimpeachable scientific analysis that establishes a causal relation between pollution and adverse health effects.

Notes

1. CEQ, *The President's 1971 Environmental Program* (1971), p. 26.

2. David A. Lynn, "Air Pollution," in W.W. Murdoch, ed., *Environment: Resources, Pollution and Society,* 2d ed. (Sunderland, Mass.: Sinauer Associates, 1975), p. 234.

3. *Washington Post,* July 4, 1977, p. A1.

4. Walter W. Heck, "Air Pollution and the Future of Agricultural Production," in American Chemical Society, *Air Pollution Damage to Vegetables* (Washington, D.C., 1973), p. 128.

5. CEQ, *1971 Environmental Program,* p. 26.

6. EPA, *The Economics of Clean Air* (1972), table 1-2.

7. *New York Times,* March 28, 1978, p. 53.

8. Comptroller General of the U.S., *U.S. Coal Development—Promises, Uncertainties* (1977), p. 2.7.

9. *Washington Post,* July 4, 1977, p. A1.

10. U.S. House of Representatives, Committee on Interstate and Foreign Commerce, Subcommittee on Health and the Environment, hearings, *Clean Air Act Amendments—1975, Part 2,* 94th Cong., 1st sess., 1975, p. 754.

11. EPA, Office of Planning and Evaluation, *Economic and Financial Impacts of Federal Air and Water Pollution Controls of the Electric Utility Industry,* EPA-230/2-76-013 (1976). The statement that the utility industry's expenses are the largest of any industry comes from CEQ, *Environmental Quality—1976* (1976), p. 160.

12. Chase Econometric Associates, "The Macroeconomic Impacts of Federal Pollution Control Programs: 1976 Assessment" (Washington, D.C., 1976), p. 18.

13. William W. Lowrance, *Of Acceptable Risk* (Los Altos, Calif.: William Kaufman, 1976), p. 8.

14. Francis E. Rourke, *Bureaucracy, Politics, and Public Policy,* 2d ed. (Boston: Little, Brown, 1976), p. 134.

15. Ibid.

16. U.S. Senate, Committee on Public Works, Subcommittee on Air and Water Pollution, hearings, *Air Pollution—1967 (Air Quality Act)—Part 3,* 90th Cong., 1st sess., 1967, p. 1154.

17. Cited by W. Clark Wescoe, "A Producer's Viewpoint," in NAS, *How Safe Is Safe?* (1974), p. 28.

18. The major exception to this generalization involves the mandatory testing of prescription drugs.

19. Cited in Garrett Hardin, *Exploring New Ethics for Survival* (New York: Penguin Books, 1973), pp. 61–62.

20. Ibid., p. 59. It should be noted that Hardin says that people accused of crimes should be considered innocent until proven guilty.

21. Peter Barton Hutt, "A Regulator's Viewpoint," in NAS, *How Safe Is Safe?* p. 121.

22. Ibid., pp. 121–122.

23. Alvin M. Weinberg, "Science and Trans-Science," *Minerva* 10 (April 1972):209.

24. Lowrance, *Of Acceptable Risk,* p. 79ff. Much of the discussion that follows relies on Lowrance.

25. NAS, National Research Council, *The Effect on Population of Exposure to Low Levels of Ionizing Radiation* (1972), p. 113. Cited by Lowrance, *Of Acceptable Risk,* p. 85.

26. Senate, *Air Pollution—1967,* p. 1154.

27. A British thermal unit (Btu) is the amount of heat required to raise the temperature of one pound of water one degree Fahrenheit. Because different

coals have different heat valves per pound of coal (measured in Btu), sulfur restrictions are frequently set on the basis of allowable pounds of sulfur per million Btu.

28. J. Clarence Davies III and Barbara S. Davies, *The Politics of Pollution,* 2d ed. (New York: Pegasus, 1975), pp. 185-186.

29. Theodore J. Lowi, "American Business, Public Policy, Case Studies, and Political Theory," *World Politics* 16 (July 1964):689-690. Others who rely on Lowi's framework include Robert H. Salisbury, "The Analysis of Public Policy: A Search for Theories and Roles," in A. Ranney, ed., *Political Science and Public Policy* (Chicago: Markham Publishing Co., 1968), pp. 151-175, and Randall B. Ripley and Grace A. Franklin, *Congress, the Bureaucracy, and Public Policy* (Homewood, Ill.: Dorsey Press, 1976).

30. Ripley and Franklin, *Congress,* p. 97. Much of the discussion of regulatory policy comes from Ripley and Franklin.

31. Ibid., pp. 97-98, 117.

32. William Lilley III and James C. Miller III, "The New 'Social Regulation,' " *Public Interest* 47 (Spring 1977):53-54.

33. U.S. House of Representatives, Committee on Interstate and Foreign Commerce, Subcommittee on Oversight and Investigations, hearings, *Regulatory Reform—Volume V,* 94th Cong., 2d sess., 1976, p. 9.

34. U.S. Senate, Committee on Government Operations, Committee Print, *Study on Federal Regulation, vol. II: Congressional Oversight of Regulatory Agencies,* 95th Cong., 1st sess., 1977, p. 81.

35. See U.S. House of Representatives, Committee on Appropriations, Report No. 92-1175, *Department of Agriculture—Environment and Consumer Protection Appropriation Bill, 1973,* 92d Cong., 2d sess., 1972, p. 67; "EPA Study: National Academy Set to Serve Two Masters," *Science,* no. 4152 (August 23, 1974):678; and U.S. House of Representatives, Committee on Appropriations, Report No. 93-1379, *Agriculture, Environmental and Consumer Protection Appropriation Bill, 1975,* 93d Cong., 2d sess., 1974, p. 5.

2 Assessing Risk: Scientific and Methodological Concerns

Lowrance suggests that scientists' assessments of the risks associated with pollution are the product of a four-step process.[1] First, scientists must indicate the conditions of exposure. This entails a determination of the volume of pollution that exists, the extent of a population's exposure to this pollution, and the circumstances of the exposure. For example, are urban residents confronted with large volumes of SO_2 for short periods, or is the pollutant evenly distributed among urban and rural areas?

Second, after a pollutant is identified as a hazard, the next step is to determine the nature of the hazard. Does the pollutant cause eyes to water or skin to itch, or are more serious problems, such as carcinogenesis, mutagenesis, or teratogenesis, likely?[2] The identification of adverse effects is often a complex process, because scientists often have no idea where to begin or which questions to ask. Similarly a pollutant can have detrimental consequences that are not apparent for many years; usually fifteen to twenty years pass between a person's exposure to a carcinogenic chemical and the manifestation of cancer, for example.

Third, scientists must relate exposure to effect. In doing so, they attempt to determine how conditions of exposure are related to adverse effects. They ask, for example, what reactions can be expected from short periods of high exposure or from long periods of low exposure, and what levels are especially damaging to certain segments of the population.

Once these preliminary steps are concluded, scientists can move to the most important. In what has been called the *payoff stage,* "the various consequences of exposure are summarized to yield overall estimates of risk, both to individuals and society as a whole."[3]

The results of this assessment are crucial to environmental policy making. They are likely to spark controversy because they are a prelude to regulatory action. The high probability of political controversy does not mean, however, that scientists are ill equipped to affect the outcome. In at least one respect, the scientific community can play an important role by providing credible assessments. Policy makers will be appreciative of these assessments since they will limit the basis for dissension from those subject to regulation and decrease the need for officials to defend the scientific grounds for their value judgments.

Public officials can be major beneficiaries of sound research. Accordingly it is to their advantage to foster such work by providing adequate resources—money, personnel, equipment, and consistent direction. Officials will also want

17

to ensure that competent researchers do not suffer because of the political controversies that stem from their work.

Just as political considerations can positively affect scientific research, they can also have the opposite effect. As an illustration, scientists working for regulatory agencies can find themselves under great pressure to produce results consistent with policy makers' preconceived notions about what the results should be. Regulated interests are not unaware of this possibility, as the National Research Council has pointed out:

> Research performed or commissioned by either party to what is likely to become a legal dispute is generally regarded as suspect, at least by the other party. Unfortunately it is prudent to be aware that self-interest may influence the design, conduct, or reporting of research, whether performed on behalf of the regulated or the regulator.[4]

In addition to the possibility of intentional bias, regulatory agencies frequently face congressionally mandated deadlines for administrative action. This problem is especially acute for the EPA. Every law that it administers includes some statutory deadlines, which often bear no relation to the time needed to gather and analyze relevant scientific information.[5] Under these circumstances, any scientific data, regardless of their validity, can be appealing to policy makers faced with legal deadlines or the possibility of a court suit if they miss a deadline.[6] For some administrators, compliance with arbitrary time limits can be more important than the need for reliable data. If this situation occurs, then environmental policy makers may be willing to proceed with obligatory standard setting even without scientific justification for their actions.[7]

Politicians are not alone in their ability to constrain scientists and their work. Most scientists will readily admit that there are limits to what they can do. The study of environmental degradation is a relatively new endeavor, and the fact that it involves the most complex of all systems does not facilitate research. So much is unknown and uncertain that even after meticulous examination of a particular ecosystem, it will still be difficult to characterize the consequences of pollution.[8] The EPA's scientific requirements are particularly difficult to accomplish, as the following explanation suggests:

> First, since widespread concern with the environment is relatively new and EPA's responsibilities have only recently been defined in legislation, the number of existing scientific and technical organizations with competence in addressing EPA's specific concerns is limited. Second, the range of technical and scientific issues affecting EPA decisions is unusually broad, encompassing different types of effects, different chemicals, and different industries and technologies. Third, the focus of EPA's interest changes as new data emerge and as political and economic pressures change. While such difficulties are to be expected in the first years of a new agency, they have been particularly troublesome for EPA.[9]

Although the tasks of environmental scientists are not easy, little has been said about how the limits of science affect the assessment of risk. The rest of this chapter explores two of these limits: the ability to measure pollution and the ability to relate varying levels of pollution to adverse effects.

The Measurement of Air Pollution

Both scientists and policy makers benefit from accurate measurements of air pollution. They allow policy makers to detect changes or trends in pollution levels, to evaluate the success of their control programs, and to determine if air quality standards are being achieved or violated. For researchers, precise measurements are necessary to detect environmental problems at early stages and to assess the risks of pollution. In fact, the ability to measure accurately is an indispensable condition of risk assessment. With adequate measurements, statements of risk may be subject to question; without them, such statements will be subject to considerable doubt.

Concentrations of SO_2 are usually measured in parts per million (ppm) or in micrograms per cubic meter ($\mu g/m^3$).[10] An initial problem with the ability to measure accurately such low levels involves the fact that there are many different ways to measure the pollutant. Consequently it is often impossible to compare data gathered by separate monitoring techniques.

In addition to providing low levels of comparability, many of the monitoring devices are expensive to operate, require frequent calibrations that are often difficult to make, have complex operational features, are sensitive to changes in humidity and temperature, and produce erratic results if other pollutants are present, which is usually the case for studies conducted outside the laboratory. Measurements reported in the mid-1960s were so inaccurate that statements were often attached to data reports to indicate that "values may at times be greater or less than the true sulfur dioxide concentration."[11] By the end of the 1960s, the situation was no better. The government's top air pollution official told members of one congressional subcommittee in December 1969 that systems for measuring and monitoring air pollution were in their infancy.[12]

Shortly after its creation in late 1970, the EPA designated a single procedure as a reference method for measuring SO_2. A reference method is an "authorized single protocol which describes how a sample of the pollutant should be taken and includes a description of the instruments and procedures to be used in making a measurement."[13] Reference methods are usually the result of several years' work, but in the case of those for air pollutants,

to meet the deadlines established by the Clean Air Amendments of 1970, the [Environmental Protection] Agency was compelled to produce reference methods for the [major] air pollutants in great haste.

For some of the air pollution characteristics, no satisfactory measurement methods were available at the time the amendments were enacted;
and, for other characteristics, several alternative methods were available. Not surprisingly, much controversy attended the initial publication of the air pollutant reference methods.[14]

For the measurement of SO_2, the EPA selected the West-Gaeke procedure
as the reference method because it was the method with which most researchers
wore familiar.[15] Despite its designation, the EPA has not been entirely successful
in its efforts to get state and local agencies to use it or other acceptable methods.
One survey of monitoring sites found that almost every fourth site operated by
the states relied on an unacceptable method of measurement.[16] Problems with
cost, limited resources, poor organization, and a reluctance to discard unacceptable devices all contributed to the problem.

Even when the monitoring devices are identical and closely supervised,
problems can still arise. In the late 1960s, for example, the Department of
Health, Education and Welfare (HEW) initiated a nationwide study of several air
pollutants, but HEW (and later the EPA) had major difficulties with the West-
Gaeke monitoring devices. In their examination of data samples, HEW/EPA
seemingly neglected the influence of changes in temperature, deficiencies in
instrumentation, and disparate means for collecting and analyzing the samples.[17] A retrospective appraisal of the study found that some samples underestimated actual levels of SO_2 by as much as 100 percent, while other samples
between 50 and 200 micrograms per cubic meter were rated as unusable.[18]

The continual production of disparate results made it imperative for the
EPA to develop an extensively improved monitoring system, which it started in
1972. The resulting Community Health Air Monitoring Program (CHAMP)
comprises a second-generation automatic system with approximately twenty
stationary sites throughout the country. By late 1976, however, the CHAMP's
monitoring devices were not yet validated. When the system does become fully
operational, the range of error will be approximately 15 to 20 percent.[19] For
most scientific research, such variability would be intolerable, but for the
measurement of air quality, the CHAMP network represents a significant improvement over other systems.

A second problem facing scientists is their general inability to get long-term
data on trends or changes in levels of SO_2. For most of the last twenty-five
years, there have been relatively few monitoring sites. According to the Public
Health Service, prior to 1956 there were fewer than two hundred communities
in twenty-nine states in which any measurements were recorded for all major air
pollutants. Nine of the states monitored pollution in only one city, and two
other states accounted for nearly half of all monitoring. Of all the sites, fewer
than half monitored SO_2.[20] In addition to the relative scarcity of monitoring
sites in the 1950s, one official from the Public Health Service emphasized that

the few bits of information gained were not of much value. He said in 1958 that the "present state of knowledge [about air pollution] is such that many analyses now made are never interpreted or made use of in any way."[21]

The total number of monitoring sites operated by state and local governments increased rapidly throughout the 1960s, but this rise did not ensure comparability in the measurement methods among different cities or even within the same city. It seems that the major emphasis has been on increasing the number of monitors at the expense of ensuring their accuracy and comparability. An important scientific rule summarizes the error of this approach: few observations carefully collected are more desirable than many observations routinely collected with less care.[22] The EPA has gradually come to grips with this principle because it now recognizes that state and local governments probably operate too many monitoring sites for SO_2.

Although these lower levels of government conduct more than 90 percent of all monitoring, the national government has not played an insignificant role. Federal efforts to monitor concentrations of SO_2 began in the mid-1950s when the Public Health Service created the National Air Surveillance Network (NASN). This network, which now uses the West-Gaeke technique, included approximately 50, 100, and more than 250 sites in 1966, 1968, and 1975, respectively. Most of these sites initially required manual chemical analysis, and samples were collected about twice a week. The NASN sites were originally supervised by a single office, but beginning in 1973, control was decentralized to some states and to EPA's ten regional offices throughout the country.

A second system, the Continuous Air Monitoring Program (CAMP), which was initiated in 1962, is automated and collects samples every five minutes. There are, however, only a few CAMP sites with only one monitoring device in a handful of cities. Though few in number, the CAMP stations represented the only federally operated and directed sources of air quality data for the six most important pollutants until the early 1970s. Neither the NASN nor the CAMP sites are particularly useful in producing longitudinal data because the monitoring instruments are dissimilar and have been changed since the projects began. As a result, the two systems produce results of dubious comparability to the point where less than half of the data collected in both systems are considered to be valid on a national basis.[23] Moreover since both systems are so labor intensive, the data tend to be several years old when published.

Still another problem with measurement concerns the location of monitoring sites. Ideally sites should be placed to give a representative sampling of an area's air quality. When there were relatively few sites, as in the 1960s, this objective was impossible to achieve. Many cities had only one site, and this was often located in the central business area, where concentrations of pollution tend to be relatively high. Readings at these single sites were often used as summary measures for an entire metropolitan area, a method that obviously led to inappropriate generalizations.

The experience gained over the last decade has led to the realization that such factors as the location of major polluters and local meteorological conditions must be considered before placing monitoring devices. To some degree, the EPA attempted to incorporate these considerations in the guidelines for the reference methods it issued to the states in 1971. These guidelines indicate that monitors are intended primarily to measure pollutants in the ambient air, which the EPA has defined as "that portion of the atmosphere external to buildings to which the general public has access."[24] This criterion would seem to suggest that monitoring stations be placed at ground level near points of high population concentration.

Shortly after the guidelines were issued, however, the EPA also said that monitors should be placed so that they can measure extreme values, natural background levels, the level of population exposure, and the amount of pollution that particular sources emit.[25] These supplementary purposes are not always compatible and can conflict with one another. Although it may be possible to establish a network of monitors that achieves all of these purposes, most state and local agencies tend to have other thoughts in mind when they make decisions about where to place monitors. When the National Academy of Sciences conducted a study of placement considerations in 1974, it concluded that most sites are selected on the basis of "(a) population density (high density preferred), (b) distance from major stationary sources, (c) price of real estate, (d) maintenance, and (e) protection against vandalism. Quite frequently, the last criterion has dominated the choices."[26] One other determinant of placement involves the ability to record low levels of pollution in order to comply with air quality standards. Accordingly if a station records high levels at one spot, the station can always be moved to record lower levels at another spot.

Finally problems with measuring SO_2 appear because of a lack of adequately trained personnel to operate monitoring equipment. This inadequacy is best illustrated by a test carried out in the early 1970s to ascertain the comparability of measurements made by eighteen qualified laboratory analysts. Each analyst was given a standard measure of 60 micrograms per cubic meter of SO_2, yet the analysts estimated the results to be between 0 and 130 micrograms per cubic meter.[27] Although the range of these errors is large, it is probably smaller than would be expected when unqualified technicians handle measurements. The absence of adequate numbers of qualified analysts is especially detrimental because state and local agencies now operate over eight thousand monitoring stations.

One possible means of improving manpower skills would be to ensure the existence and application of an adequate quality control program. Indeed such a program is so necessary that the CEQ believes that the most critical need of current monitoring programs is a means of ensuring the accuracy of the data collected.[28] Despite this admonition, the EPA did not have such a program at its own stations until 1974. As for state and local agencies, as late as 1975 federal guidelines on monitoring did not require that these agencies have any quality assurance program. And a year later, EPA officials told a Senate subcommittee that the agency did not know the status of state and local monitoring with respect to compliance with national guidelines on such matters as proper siting.[29]

In some cases the EPA has attempted to evaluate quality control at the state and local levels, and the results have been discouraging. In one such evaluation in Illinois and Michigan, serious deficiencies in monitoring activities were caused by poorly trained operators, inadequate laboratory support facilities, marginal plans for quality control, and mistakes in data handling and calculation.[30] Faulty data can have important short-term as well as embarrassing long-term impacts. As an example of the former situation, poor data led eight cities to announce incorrectly health advisory alerts or violations of air quality standards on a single day in 1973.[31] Announcements of alerts or violations can trigger mandatory shutdowns of industrial operations, and they can also cause undue worry to people suffering from certain respiratory ailments. Over the long term, wrong data can lead policy makers to conclude that their control programs are inadequate when in fact they are not. Of course, the opposite situation can also result from poor data—actual violations of standards can go unnoticed.

Great strides were made in the early 1970s in improving measuring devices' accuracy, reliability, and comparability, but much is left to be done. Improvements are still necessary in the areas of training, evaluation, and selection of sites. In recognition of the need for drastic improvement, both the Senate and the House of Representatives have become involved. Recent legislation on air pollution includes provisions on monitoring and measurement. In its discussion of proposed changes to existing legislation in 1977, for example, the Senate Committee on Public Works concluded that "information on the current levels of air quality, both in urban and rural areas, is inadequate."[32] At the insistence of the House of Representatives, the Clean Air Act Amendments of 1977 mandate improvement and standardization of monitoring techniques throughout the country.[33] The Congress gave the EPA one year to implement this directive. It is not surprising that there seemed to be no relation between the length of time allotted and EPA's ability to comply.

To summarize, the state of monitoring and measurement leaves much to be desired. At best, measurements of air pollutants, including SO_2, tend to be inaccurate and unreliable, and the prospects for rapid improvement do not appear promising. A recent examination of the nation's environmental monitoring network concluded that it is fragmented, poorly coordinated, inadequately funded, and characterized by weak quality control. "At present," said the Study Group on Environmental Monitoring of the National Research Council,

> there are significant gaps in environmental monitoring, as well as a proliferation of uncoordinated, inefficient, and inflexible programs that

produce data of poor or unknown quality. . . . Too often monitoring
programs are developed, not according to scientific principles, but in
hasty response to the dictates of legislation, enforcement procedures,
and the exigencies of pollution control management.[34]

The blame for these problems was placed clearly on the shoulders of the government's environmental sentry: the Environmental Protection Agency.

Relating Pollution to Adverse Effects

Measurements tell scientists how much pollution exists but not how these levels
are related to ill effects. To discern the relation between the two, scientists
usually rely on three research methods: clinical, toxicological, and epidemiological studies. Each of these methods has its own strengths and weaknesses. Paradoxically the method that relies on actual conditions produces the least tenable
cause-and-effect relationships, while the other two methods, which rely on
artificial environments, produce the strongest relationships.

Epidemiological studies attempt to compare the incidence of harm for
selected groups of individuals exposed to known pollutants with incidence
rates for a similar group of people who have either not been exposed or have
been exposed at much lower levels.[35] Accordingly such studies employ classic
research designs with control and treatment groups. Once the groups have been
selected, researchers try to detect changes in the health of respondents, usually
through the use of questionnaires or through examination of hospital records
and death certificates. Since the epidemiologist's laboratory is the environment
itself, all possible relevant factors are available for study. This trait is probably
the major strength of epidemiological research. In addition, such research is
especially meaningful for the study of air pollution effects "since the popula-
tions studied can cover the whole range of ages and levels of good health or ill
health, and environmental factors are studied within the context of a broad
range of factors which affect health."[36]

At first glance epidemiological studies seem to be ideal instruments for
relating exposure to pollution with effects, since they can provide results of
immense value. Yet well-conducted epidemiological studies are almost impossible to achieve; they usually present researchers with every conceivable methodological sin. For example, a prime tenet of a good research design requires that
subjects be assigned randomly to control and treatment groups. This practice
helps to ensure that both groups are as similar as possible before the study begins
and that the differences between the groups at the end of the study can be
attributed to the test variable.

Random assignment is rarely possible in epidemiological studies of air
pollution because researchers must find a community with the pollutants they

wish to test and then find volunteers who are willing to provide extensive personal and medical histories, report on all changes in health, and remain in the community during the duration of the study. The nonrandom selection of participants and research sites means that neither is representative of the population or pollution conditions, and this reduces the ability to generalize beyond those involved in the study.

Epidemiological designs are also beset with problems of measurement and data analysis. Reliable measurements of ambient air quality are not easily obtained, and epidemiological studies are highly dependent on precise measurements of outdoor air quality. Consequently researchers face great difficulty in establishing the lowest levels of pollution that cause harm. Another problem related to measurement is the fact that stationary monitoring sites record levels of pollution only in the vicinity of the sites. It is highly unlikely, however, that a study's participants will spend much of their time around the sites. The levels of air pollution to which a person is exposed will be far different from that at monitoring sites due to differences in meteorologic conditions, personal habits, amount of time spent indoors, and so forth.

Attempts to link exposure to harm are further confounded because of the exceedingly large number of possible covariates, including age, sex, race, occupation, education, smoking, general health, and relative immunity to certain diseases. As the number of covariates increases, efforts to relate exposure to harm become more difficult. If a researcher is not aware of all covariates or ignores them in the analysis, the results will be of questionable value.

Another problem relates to the difficulty in handling the interactions among pollutants. If the environment had only one adverse pollutant, then it would be relatively easy to discover this pollutant's effects. If many environmental pollutants are present simultaneously, however, their interaction must be considered. These reactions can be indifferent (the presence of other pollutants does not change the effect of the single most active pollutant), additive (the total effect is equal to the sum of the individual pollutants), antagonistic (the effect is less than the sum of the individual pollutants), or synergistic (the effect is greater than the sum of the individual pollutants).[37] Of the four possibilities, the last is probably the most troublesome and the source of greatest concern. With synergistic reactions, the effect of each pollutant is multiplied, often by some unknown factor. As an illustration, smokers exposed to asbestos are substantially more likely to have a higher incidence of lung disease and certain cancers than are nonsmokers.

The possibility of synergistic reactions can make an epidemiologist's work exceedingly difficult, especially if he is trying to establish the probability of risk associated with environmental pollutants. Under ideal laboratory conditions, a researcher would have some control over synergistic reactions, but outdoors a scientist can neither control the possible reactions nor even know the possible combinations. For example, an epidemiologist would like to know if the adverse effects of SO_2 are enhanced in the presence of hydrocarbons and particulates or such esoteric air pollutants as benzopyrene or vinyl chloride, and if so, then under what climatic, geographic, or atmospheric conditions. In spite of these substantial problems, the probabilities of risk cannot be ascertained fully unless possible synergistic effects are explained. To do less might lead to two serious errors: identifying SO_2 as the sole culprit while exonerating other equally culpable criminals or understating the probabilities of risk.

Finally because epidemiological studies are so expensive, they are better suited for study of acute as opposed to chronic effects. Acute effects are those that occur shortly after exposure to a substance. For certain pollutants, these effects can include coughing, shortness of breath, and irritation of the eyes or skin. In contrast, chronic effects are evident only after an extended period of time. As such, they "are muted, often invisible, and usually delayed responses to doses which do not produce overt, harmful manifestations."[38]

The relation between acute and chronic effects is not always clear since exposure to a harmful substance over a short period may have no immediate effect but contribute to disastrous long-term consequences. As a result, the absence of acute effects cannot be taken as an indication that a pollutant is harmless. Thus research directed at determining probabilities of risk should be concerned with both acute and chronic effects.

The problem with epidemiological studies is that the latency period of some diseases is so long that these studies are entirely impractical when trying to discover certain chronic effects. Cancer provides the best illustration of a chronic effect since its latency period can exceed thirty years. Other examples of diseases with long latencies include emphysema, pulmonary fibrosis, and cirrhosis of the liver.

In sum, epidemiological studies are imperative because the natural environment serves as the scientists' laboratory. Such studies provide the ideal setting for research but are beset with many problems that reduce their usefulness. Scientists do have, however, two other methods to overcome some of these difficulties.

Clinical studies rely on the use of healthy volunteers in a laboratory setting. This procedure provides scientists with precise control over the conditions of exposure, which means they have accurate knowledge of doses and length of exposure. Thus researchers can more easily establish a cause-and-effect relationship if they know that repeated experiments produce similar results. In addition to controlling the conditions of exposure, clinical studies also allow researchers to control for or eliminate the effects of covariates. A further advantage of clinical studies is their ability to avoid reliance on imprecise measurements of ambient air quality. Because clinical analysis takes place in laboratories, researchers do not have to depend on untrained technicians to collect information on levels of air pollution. This means, of course, that researchers have much greater control over the design of measuring instruments, an important contribution to competent research.

The advantages of clinical studies tend to be offset by the method's critical limitations. First, and most important, a laboratory setting cannot duplicate the ambient air. This problem can lead to misleading results by indicating incorrect levels at which harm occurs and by ignoring possible synergistic reactions. Second, ethical considerations exclude the use of children, the aged, or people who are already ill or highly susceptible to harm. Only healthy adults can be the subjects of clinical tests, but such people are hardly representative of the larger population. It can also be difficult to find a sufficient number of volunteers because dosages in clinical tests are usually continued to the point where effects are easily observed. If such effects are chronic or irreversible, it is unethical to conduct clinical studies. Third, the costs and limited duration of clinical studies make them of little use in studying chronic effects. All these difficulties mean that clinical research cannot be used as the sole means of assessing the risks associated with any pollutant.

Animal or toxicological studies seem to offer several means of overcoming the inherent limitations of both clinical and epidemiological studies. Animal studies allow scientists to use powerful research designs with random assignment of subjects to test and control groups and several other features that improve their validity. When animals are used, ethical considerations are less important, human suffering is avoided, and chronic effects can be studied. Also, because many animals have short breeding periods and life spans, the animals can be used to test for carcinogenesis, mutagenesis, and teratogenesis.

The absence of legal and ethical considerations allows researchers to test dosages and periods of exposure that could not be used on humans. This ability is important because it allows researchers to give huge doses over short periods of time, and the "higher the exposure of a subject (irrespective of exposure duration), the greater the likelihood that disease will be contracted by either animal or man."[39] This high-dose–short-exposure relationship means that some harmful effects with an extended latency period in humans will appear in some animals in a year or two.

Like their laboratory counterparts, animal studies are not without their problems. Animals exposed to air pollution can show gross effects such as death or lung damage, but they cannot express such things as irritation or discomfort. Animal and clinical studies share similarities in that they both start with healthy subjects, and neither duplicates the outdoor environment, is especially amenable to analysis of synergistic or antagonistic effects, or is dependent on unreliable measures of pollution.

A major problem with animal studies is the difficulty in generalizing to humans. Lowrance states the problem well: "Adverse reaction in animals does not prove adverse effect in man, and lack of reaction in animals does not prove that man will not be affected."[40] As an illustration, he observes that thalidomide causes birth defects in monkeys and rabbits but not in rats. The linkage between ill effects in animals and similar effects in humans is especially weak for pulmonary effects "in view of the relatively marked difference in respiratory tract anatomy and physiology between man and most laboratory animals."[41] This fact lessens the appeal of animal studies for scientists examining the effects of air pollution inasmuch as these effects are frequently related to problems with heart and lungs.

This discussion of the three major methods for linking exposure to air pollution with adverse effects demonstrates the difficulties that confront scientists. Each method has its own advantages, yet each has its drawbacks; none is satisfactory by itself, so scientists must rely on all three.

Once researchers have the results of many tests, their final scientific responbility is to synthesize these results, resolve conflicts to the extent possible, and then estimate the overall risk. Private researchers usually prefer that government scientists complete these tasks, since the work often requires that subjective judgments be made. Thus governmental scientists usually find that it is their job to summarize the data in order to specify the levels of pollution and lengths of exposure at which harmful effects occur. Such scientists may also be expected to indicate the existence of any threshold levels, the point below which no adverse health effects are found. Estimating these levels can be difficult, particularly since some uncertainty exists as to whether it is possible to define threshold limits for any air pollutant. Such terms as *no effect* or *below threshold* lack precise scientific meanings. Even if some agreement did exist on the definition of these terms, the definition would be subject to change as a result of new tests, new measurement techniques, or old tests that are repeated under new conditions.

That scientists cannot prescribe exact threshold levels does not mean that they have abdicated their responsibility. To the contrary, a researcher has reached the limit of his scientific responsibility when he has estimated the probability and severity of harm. Those people who go beyond this responsibility by arguing for or against certain levels of air pollution shed their scientific mantle and enter an arena in which professional objectivity is less important than an ability to balance satisfactorily a pollutant's medical risks and its social and economic benefits. This is a further reason that scientists outside the government like to avoid the final step in risk assessment.

Conclusions

The measurement of pollution and the assessment of risk are integral to the determination of acceptable levels of air quality. Unfortunately in both areas scientists confront a number of major limitations that diminish the prospects that they will play an important or determinant role in policy making. For policy makers, the limitations lead to at least two consequences. First, an inadequate understanding of the effects of SO_2 increases the likelihood that inappropriate decisions will be reached. Second, equivocal scientific findings allow decision makers to select whatever results suit their interests or to neglect scientific assessments altogether.

Notes

1. William W. Lowrance, *Of Acceptable Risk* (Los Altos, Calif.: William Kaufman, 1976), pp. 19–41.

2. Carcinogenesis, says Lowrance, "is the induction of runaway growth of the cells of some part of the body"; mutagenesis is alteration of the body's genetic material, and teratogenesis is modification of a baby's development in the womb. See ibid., pp. 26–27.

3. Ibid., p. 41.

4. National Research Council, *Research and Development in the Environmental Protection Agency* (Washington, D.C.: NAS, 1977), p. 81.

5. National Research Council, *Decision Making in the Environmental Protection Agency* (Washington, D.C.: NAS, 1977), pp. 66–71.

6. Environmental groups such as the Sierra Club and Natural Resources Defense Council have frequently brought suit against the EPA in efforts to compel compliance with statutory deadlines.

7. For a useful discussion of policy makers' tendency to discourage sound scientific analysis, see Edward J. Berger, Jr., "Regulation and Health: How Solid Is Our Foundation?" *Environmental Law Reporter* 5 (1975):50179–50187.

8. National Research Council, *Perspectives on Technical Information for Environmental Protection* (Washington, D.C.: NAS, 1977), p. 15.

9. National Research Council, *Decision Making in the EPA*, p. 59.

10. One part per million equals approximately 2,618 micrograms per cubic meter.

11. U.S. Senate, Committee on Public Works, Subcommittee on Air and Water Pollution, hearings, *Air Pollution—1967 (Air Quality Act), Part 4,* 90th Cong., 1st sess., 1967, p. 2533.

12. U.S. House of Representatives, Committee on Interstate and Foreign Commerce, Subcommittee on Public Health and Welfare, hearings, *Air Pollution Control and Solid Waste Recycling, Part 1,* 91st Cong., 1st and 2d sess., 1969–1970, p. 78.

13. National Research Council, *Environmental Monitoring* (Washington, D.C.: NAS, 1977), p. 66.

14. Ibid., pp. 66–67.

15. *Federal Register* 36 (April 30, 1971):8187–8190. These pages describe the method and its application.

16. CEQ, *Environmental Quality—1975* (1975), p. 336.

17. U.S. House of Representatives, Committee on Science and Technology, Subcommittee on Special Studies, Investigations and Oversight and Subcommit-

tee on the Environment and the Atmosphere, Committee Print, *The Environmental Protection Agency's Research Program,* 94th Cong., 2d sess., 1976, pp. 31–37. For similar findings, see National Research Council, *Environmental Monitoring,* pp. 40–41.

18. House, *EPA's Research Program,* pp. 12, 36.

19. Ibid., p. 44

20. HEW, PHS, *Proceedings: National Conference on Air Pollution (1958)* (1959), p. 100.

21. Ibid.

22. For the emphasis on the number of monitoring sites, see CEQ, *Environmental Quality–1975,* p. 309, and EPA, *Progress in the Prevention and Control of Air Pollution in 1973* (1974), pp. 51–53. The scientific rule is cited in National Research Council, *Environmental Monitoring,* p. viii.

23. U.S. House of Representatives, Committee on Science and Technology, Subcommittee on the Environment and the Atmosphere, hearings, *The Costs and Effects of Chronic Exposure to Low-level Pollutants in the Environment,* 94th Cong., 1st sess., 1975, p. 872.

24. *Federal Register* 36 (April 30, 1971):8186–8201.

25. National Research Council, *Air Quality and Automobile Emission Control,* vol. III: *The Relationship of Emissions to Ambient Air Quality,* a report by the Coordinating Committee on Air Quality Studies, NAS, National Academy of Engineering for the U.S. Senate, Committee on Public Works, Committee Serial No. 93-24, 93d Cong., 2d sess., 1974, p. 46.

26. Ibid., p. 42.

27. House, *Costs and Effects of Chronic Exposure,* p. 841.

28. CEQ, *Environmental Quality–1975,* p. 336.

29. See House, *Costs and Effects of Chronic Exposure,* p. 873, and U.S. Senate, Committee on Appropriations, hearings, *Department of Housing and Urban Development, and Certain Independent Agencies Appropriations, Fiscal Year 1977, Part 3,* 94th Cong., 2d sess., 1976, p. 798.

30. CEQ, *Environmental Quality–1975,* pp. 336–337.

31. National Research Council, *Environmental Monitoring,* p. 31.

32. U.S. Senate, Committee on Environment and Public Works, Report No. 95-127, *Clean Air Act Amendments of 1977,* 95th Cong., 1st sess., 1977, p. 17.

33. The Clean Air Act, PL 95-95, sec. 319, August 7, 1977. A month after the law's enactment, members of a House subcommittee conducted three days of hearings to review monitoring activities. See U.S. House of Representatives, Committee on Science and Technology, Subcommittee on the Environment and the Atmosphere, hearings, *Environmental Monitoring,* 95th Cong., 1st sess., 1977.

34. National Research Council, *Environmental Monitoring,* p. 12.

35. CEQ, *Environmental Quality–1975,* p. 26.

36. EPA, Science Advisory Board, "Review of the CHESS Program" (mimeographed report, March 14, 1975), p. 1.

37. National Research Council, *Air Quality and Automobile Emission Control,* vol. II: *Health Effects of Air Pollutants,* p. 485.

38. U.S. House of Representatives, Committee on Interstate and Foreign Commerce, Report No. 94-1175, *Clean Air Act Amendments of 1976,* 94th Cong., 2d sess., 1976, p. 97.

39. CEQ, *Environmental Quality–1975,* pp. 29–30.

40. Lowrance, *Of Acceptable Risk,* p. 64.

41. National Research Council, *Air Quality and Stationary Source Emission Control,* a report by the Commission on Natural Resources, NAS for the U.S. Senate, Committee on Public Works, Committee Serial 94-4, 94th Cong., 1st sess., 1975, pp. 14–15.

3 Initial Efforts to Define Acceptable Levels

In October 1948, the residents of Donora, Pennsylvania, found themselves surrounded by a "motionless clot of smoke," and visibility was near zero. As one observer recalled, "The air began to have a sickening smell, almost a taste. It was the bittersweet reek of sulfur dioxide."[1] The wind had stopped blowing, and the town was engulfed in high levels of pollution for nearly a week. Twenty people died as a result of the pollution, and thousands of others suffered from nausea, vomiting, diarrhea, sore throats, and burning of the eyes.

The incident received much publicity, but the federal government did not pass legislation addressed to the problems of air pollution until 1955. Public Law 84-159, which had no title, authorized the annual expenditure of $5 million for a five-year period so that the federal government could conduct and contract for research, provide such technical services as training, and give some financial assistance to state and local agencies concerned with air pollution. In addition, the law contained several sections that could be interpreted to allow HEW's Public Health Service (PHS) to conduct and publish research relating to safe levels of air quality.[a] Both the House and Senate made it clear that the law was not an attempt to impose standards of purity on the states, but the law did authorize the surgeon general and the Public Health Service to "collect and disseminate information relating to air pollution and the prevention and abatement thereof" and to make available to air pollution agencies "the results of surveys, studies, investigations, research and experiments relating to air pollution."[2] The surgeon general was also authorized to prepare and publish reports "as he may consider desirable" or in response to requests from state or local governments "concerning any specific problem of air pollution confronting such state or local government air pollution control agency with a view to recommending a solution of such problem."[3]

The PHS used this authority and some of the money that the law authorized to initiate its research into the effects of SO_2 and sulfuric acid on man and animals. Thus the PHS did not start a sustained research program on sulfur oxides until seven years after the Donora incident. Even then, little of this research was directed at defining acceptable levels of air quality.

---

[a] Although PL 84-159 gave primary responsibility to HEW, the Public Health Service, which was part of HEW, and its divisions had the actual working responsibility for conducting the federal government's research on air pollution. Thus HEW and the PHS are used interchangeably.

The late start in studying the effects of air pollution can be traced to the PHS's unwillingness to become involved with the problems of abatement and control. The health service apparently realized that any efforts to determine adverse levels of air pollution would lead to controversy, especially if it would have a role in recommending air quality or emission standards. Regulation and enforcement were alien to the scientific researchers on PHS's professional staff. In addition, President Dwight Eisenhower was not interested in involving the federal government in what he considered to be a state and local function. As far as Eisenhower and the PHS were concerned, the federal government had no responsibility for defining acceptable levels of air pollution. This was a matter best left to state or local governments, even though certain economies of scale would be available if the federal government assumed the responsibility.

Despite Eisenhower's belief that the federal government should not have a role in defining clean air, the PHS was almost put into a position of sure controversy in the late 1950s when several congressmen tried to require the agency to set acceptable levels of pollutants from automobiles. The PHS resisted the efforts as much from its desire to avoid controversy as from its inability "scientifically and technically to ascertain the health effects of motor vehicle emissions."[4] There were at least two major reasons for the absence of adequate scientific information. First, scientists' ability to relate varying levels of air pollution to adverse effects was relatively undeveloped. For the most part, adequate measuring devices were unavailable, few scientists were working on the problem, and much of the work that was done was crude. Second, and perhaps of greater importance, research on air pollution had meager support from either the government or from most of the population. Few if any opinion polls exist to indicate what the public thought about air pollution in the 1950s. The absence of such polls probably reflects the fact that pollsters did not think air pollution to be sufficiently important to justify a survey. From the federal government's perspective, Congress never provided the PHS with sufficient funds to match authorized spending levels. Congress had authorized the annual expenditure of $5 million from 1956 to 1960, but the actual appropriations ranged from less than $2 million in 1956 to $4 million for fiscal year 1958.

By the early 1960s, the PHS gradually began to increase its activities in the air pollution field, at least partially as a result of the new administration in Washington. Only a month after he had taken office, President John F. Kennedy indicated his dissatisfaction with the federal government's existing air pollution efforts, but his response to the problem was largely symbolic. Kennedy called for an "effective Federal air pollution control program now."[5] The reaction from the PHS was to plan for future research and to develop national goals.

While the PHS believed that it was expanding its functions and responsibilities, its efforts during Kennedy's first year in office were apparently unsatisfactory. By early 1962 the president submitted a bill to Congress that mandated an expanded governmental research effort. The bill also provided for federal grants so that state and local agencies could initiate, develop, or improve their air pollution programs. The Congress did not grant these changes; instead it passed a two-year extension of the existing law.

The setback did not deter the president's efforts to enlarge the federal role in pollution control. In a message to the Congress in early 1963, Kennedy again recommended that the PHS be allowed to "engage in a more intensive research program permitting full investigation of the causes, effects, and control of air pollution." Of greater significance was his request that the PHS be granted the authority to expand its role in enforcement and abatement.

Kennedy's support was important, but it was not sufficient to diffuse the controversy over an expanded federal role. Most industrial groups favored minimal federal involvement, while such groups as the American Municipal Association and the U.S. Conference of Mayors supported a larger role. The Congress eventually passed air quality legislation that was the product of extensive compromise and incremental decision making. The Clean Air Act of 1963 reemphasized that "the prevention and control of air pollution at its source is the primary responsibility of States and local governments."[6] Nonetheless the act did give HEW some limited enforcement authority in cases involving interstate pollution, but then only in response to a state's request.

Although the most important issue in the debates on the legislation focused on the federal role in enforcement, the Congress was also concerned about attempts to link air pollution to ill health and property damage. Since one purpose of the new law would be to reduce levels of pollution to the point where no harm to health or welfare occurred, it would be necessary to ascertain the meaning of clean air. Of the seven separate bills that the House and Senate considered in 1963, six included some provision for publication of air quality criteria. These criteria, however, are not criteria in the usual sense of being standards or established rules. Air quality criteria are supposed to be neutral, descriptive summaries of existing knowledge about air pollutants and their effects on man, animals, buildings, and so forth. These criteria represent the product of scientific investigation in that they serve as an assessment of risk. This assessment usually does not include evaluative judgments about what levels of air pollution are good or bad. Thus air quality criteria are meant to serve as a prelude to air quality standards, which represent value-laden, political decisions about acceptable levels of risk.

Air quality criteria are also not meant to be directive or prescriptive (and standards are), but when the distinction is not made, scientists can find themselves making political decisions about acceptable levels of air quality. As an illustration, one bill introduced in the House in 1963 authorized the secretary of HEW to "recommend to air pollution control agencies . . . after such research as he determines to be necessary, such criteria of air quality as in his judgment may be necessary to protect the public health and welfare."[7] Such a responsibility was outside the realm of science. The proposal called for the secretary to make value judgments about the kinds of people to be protected and the kinds of health problems from which people would have to be protected.

In testimony before the House Subcommittee on Public Health and Safety, several organizations favored the development of criteria, but spokesmen from the National Association of Manufacturers and the U.S. Chamber of Commerce opposed any federal role in compiling criteria in the belief that this job belonged to the states. Both groups feared that any federally developed criteria might be imposed upon the states, and this position paralleled similar arguments against any federal expansion into the area of enforcement or abatement.

The opposition to federal development of all criteria reflected the fact that the appropriate distinction had not been made. Polluters had little to fear if the federal government merely summarized existing knowledge about air pollution and then left the definition of acceptability to the states. Nonetheless, some groups were clearly apprehensive about any federal role, regardless of how limited it might be. The Manufacturing Chemists Association and the American Iron and Steel Institute, among others, contended that existing levels of air pollution did not justify much concern, let alone air quality criteria. "As a matter of fact," said a spokesman from the steel industry, "it is only common sense to recognize that there is less air pollution in the aggregate today than there was 30 years ago, 10 years ago, 5 years ago, or even last year."[8] In view of this progress, it was argued, any federal activity was unnecessary.

Despite such opposition to federal involvement, the House-passed version of the Clean Air Act did not make the distinction between criteria and standards. And when the Senate's Special Subcommittee on Air and Water Pollution began its consideration of the bill, it also failed to make the proper distinction. Several of the bills introduced in the Senate had sections on criteria that were identical to those in the House. The sole purpose of one of them would be to require the secretary of HEW to conduct research leading to the publication of criteria. Publication would take place after a determination had been made "that there is substantial scientific agreement that a particular air pollution agent (or combination of agents), ... produces effects harmful to the public health or general welfare."[9] Here again, the secretary of HEW would be required to make a judgment about the significance and acceptability of various levels of harm. The bill's sponsor did not argue that the lack of criteria impeded the public's understanding of air pollution. Instead he suggested that the problem was the "lack of generally established and published *standards* to show what the harmful effect of various air pollutants are and what levels of air pollution in each case are harmful and what its effects are."[10]

With strong support from members of both the House and the Senate, as well as from HEW, there was little doubt that the legislation would incorporate some requirement for air quality criteria. The only remaining issue was whether these criteria would be summaries of scientific data, as criteria are supposed to be, or whether they would be prescriptive norms. The Clean Air Act of 1963 included both choices and maintained the confusion. The law instructed the secretary of HEW to "compile and publish criteria reflecting accurately the latest scientific knowledge useful in indicating the kind and extent of such effects which may be expected from the presence of such air pollution agent (or combination of agents) in the air in varying quantities." These criteria could be revised "whenever necessary to reflect accurately developing scientific knowledge."[11] The criteria would be developed for informational purposes only, but the law also authorized the secretary to recommend to state and local pollution agencies "such criteria of air quality as in his judgment may be necessary to protect the public health and welfare." Consequently the law gave two incompatible meanings to criteria. In the first instance, *criteria* would be descriptive summaries, and in the second, the term would be construed to be prescriptive recommendations concerning acceptable levels of air quality.[12] This inconsistency later proved to be both crucial and controversial.

The requirement that HEW publish criteria was a significant step for the federal government. Without a summary of scientific information on air pollution, it would be impossible for the states to establish credible air quality standards. Polluters could correctly claim that any standards or enforcement actions would be premature in the absence of a precise understanding of the relation between air pollution and adverse effects. In contrast, valid criteria would put control programs on a stronger footing by delineating the levels at which air pollution causes harm. Now that the law required the publication of criteria, state and local agencies might choose to delay enforcement actions until criteria became available. This situation meant, however, that the pace of the states' abatement programs would be almost entirely dependent on HEW's activity and the speed with which the agency published criteria. In short, a federal role in compiling and publishing criteria meant that the states would look increasingly to Washington before they selected acceptable levels of air pollution, especially since HEW was now in a position to recommend levels to the states. Despite the law's admonition that the criteria should be used for informational purposes only, such criteria would carry the federal government's imprimatur.

Other than establishing an important role for the federal government in identifying harmful pollutants, the Clean Air Act of 1963 did little to establish a meaningful federal role in the abatement of air pollution. The law allowed the federal government to collect and summarize knowledge about air pollution and health. Once HEW demonstrated a link between air pollution and ill effects to the public's health or welfare, the agency was authorized to notify the states, which could either utilize or neglect the information. This meant that HEW would identify the harmful effects of air pollution and then do virtually nothing about them. The law did allow HEW to develop technologies for the control or prevention of air pollution, but once developed, there was no way for HEW to compel their use.

In a few precisely defined situations, the federal government could initiate enforcement actions, but the procedure was lengthy, unwieldy, and highly protective of states' rights. In these few cases, HEW could start legal action if a series of prior conferences and public hearings failed to secure satisfactory compliance. Endangerment of the public's health or welfare was not to be the sole criterion for a court's judgment in these federally initiated cases. One part of the Clean Air Act instructed the courts to give "due consideration to the practicability of complying with such standards as may be applicable and to the physical and economic feasibility of securing abatement of any pollution proved ... as the public interest and the equities of the case may require."[13] This provision ensured that the courts would balance the desirability of clean air against other social objectives, as well as polluters' economic willingness to comply with environmental restrictions. Given the difficulties in quantifying the benefits of cleaner air, such a balancing process might serve as an effective deterrent to legal action.

Developing Air Quality Criteria

Congressional efforts to stimulate publication of air quality criteria were commendable and certainly necessary, but in requiring HEW to produce the criteria, Congress provided the agency with minimal guidance. As is typical of other regulatory policies, the Congress sketched the primary purpose of the Clean Air Act in broad terms. The law was intended "to protect the Nation's air resources so as to promote the public health and welfare and the productive capacity of its population."[14] The Congress gave little direction about what this phrase was supposed to mean, and this vagueness was typical of other sections in the law as well. For example, the act required that HEW conduct research on "the harmful effects on the health or welfare of persons by the various known air pollution agents (or combination of agents)." The agency was also supposed to publish criteria for air pollutants that produced "effects harmful to the health or welfare of persons." Unfortunately for HEW, no section of the law provided any indication of the kinds of harm that should be considered, a definition of health, any indication as to whose health should be protected or under what circumstances, or any specific guidelines that could be used as the basis of HEW's recommendations to the states.

From HEW's perspective, greater clarity was desirable, but legislation with broad purposes serves several functions for the Congress. First, broadly defined goals help to establish consensus and prevent legislators from becoming entangled in controversy. Faced with the prospect of controversy likely to be associated with regulation, legislators can pass vague statutes and thereby pass the controversy to administrative agencies, which must implement the statute. This procedure allows legislators to favor the public interest (or other desirable goals) without having to decide how this interest will be achieved.

Second, when laws propose goals like justice or democracy, these words do not call to mind a single meaning; rather they evoke an entire set of favorable attitudes and impressions. Murray Edelman believes that such symbolic words stimulate a feeling of reassurance and well-being.[15] Seen in this light, clean air laws are a symbol of the government's concern for its people and a commitment to a high quality of life. To appreciate the symbolic significance of the Clean Air Act, one needs only to realize that a more appropriate title would have been the Dirty Air Act.

Third, both Edelman and Theodore Lowi have argued that vague laws encourage the participation of interest groups in the implementation of policy. Broadly defined purposes, says Edelman, allow for "differing interpretations of the same language with different authorities, changing times, altered conditions, and varying group interests."[16] Lowi believes that vague laws require delegation of authority to bureaucrats, which in turn ensures the continuation of an administrative rhetoric that establishes "the ideal that every decision can be bargained."[17] Such bargaining allows interest groups to press for definitions or regulations that are favorable to their interests. Should interest groups believe that they have not received a fair hearing or that an administrative decision will affect them adversely, they can appeal to the Congress on the grounds that the agency's actions are inconsistent with congressional intent, whatever that might be. If one accepts this explanation, then it is likely that the confusion surrounding Congress's choice of words on criteria and standards was intentional.

In addition to the problem of vagueness, the Clean Air Act did not give HEW any indication of how or where to begin its efforts to develop and publish air quality criteria. These were important issues since HEW's choice of methods would influence the credibility of its efforts, while the selection of specific pollutants for study would affect the direction of the states' control programs. A number of considerations influenced the PHS's choice of pollutants for initial attention. While the PHS did not favor an emphasis on any one category of pollutants, more research had been done on sulfur oxides than on any other category, making them a good starting point. In addition, sulfur oxides were the most pervasive and detrimental of all major pollutants and the most susceptible to control in the 1960s. Unlike such pollutants as hydrocarbons and carbon monoxide, which are primarily produced by millions of automobiles, most sulfur oxides emanate from a substantially smaller number of stationary sources. And, unlike automobiles, which can be controlled only by recourse to out-of-state manufacturers, state and local governments can control stationary sources. If the PHS could publish criteria on sulfur oxides, local air pollution agencies could set air quality and emission standards applicable to the major sources of SO_2.

Besides these practical considerations, the Senate's discussion of the clean air legislation also may have influenced the PHS's choice of pollutants. Senator Edmund Muskie, who was chairman of the Senate subcommittee with jurisdiction over the legislation, had recognized SO_2 as a major pollutant, and his committee's report on the legislation reflected the concern: "It is the intention of the committee that both the problems of sulfur compounds and asthma and pulmonary diseases be pursued in greater depth, and that the efforts for the development of methods to alleviate these problems be intensified as rapidly as possible."[18]

Although the bill that passed in the House had not contained any provision requiring a specific focus on sulfur oxides, the House accepted the Senate's recommendation that HEW look at ways to reduce the emissions of sulfur oxides. In sum, both practical and political considerations made sulfur oxides an ideal candidate for the first criteria.[19]

The PHS had no illusions that publication of a criteria document on sulfur oxides would be easy. From a scientific viewpoint, ideal criteria would summarize accurately the latest scientific knowledge about how sulfur oxides harmed the public's health and welfare and specify the levels and time periods over which this harm occurred. In order to produce such a document, hundreds of clinical, toxicological, and epidemiological studies would be examined and evaluated, new studies would be initiated to fill gaps in existing knowledge, decisions would be made on the accuracy and validity of all studies, findings generated by different research methods would be reconciled, and final judgments would be made in the face of inconclusive or inadequate data.

In addition to these scientific and methodological concerns, however, the PHS also would have to convince both advocates and critics of environmental programs that the criteria accurately identified and assessed the adverse effects of air pollution. The ability to handle this responsibility rested to a large extent on the credibility of the PHS and its air pollution officials. If this reputation was impugned in any way, then the acceptance of and response to any criteria would be jeopardized. In the years after its creation, the PHS had developed a reputation for expertise and, therefore, credibility, but this expertise had rarely been brought to bear on problems with the potential social and economic implications of the criteria. Notwithstanding this reputation, the PHS would have to persuade many important political actors of the criteria's worth and accuracy.

Well-organized labor and industrial groups that would be concerned with criteria on sulfur oxides included the United Mine Workers, the National Coal Association, the American Petroleum Institute, the Edison Electric Institute (representing the nation's privately owned electric utilities), and the National Coal Policy Conference. This last group was formed in the early 1950s in an effort to unite all major coal owners, unions, transporters, and users. According to one analysis, it served as an alliance of diverse interests intending "to fend off any environmental legislation that might jeopardize their respective interests."[20]

The PHS also had to consider potential pressures from key congressmen. In both the House and the Senate, the committee chairmen with jurisdiction over all legislative proposals on air pollution came from West Virginia, the nation's second-largest coal-producing state. Harley O. Staggers chaired the House Committee on Interstate and Foreign Commerce (after 1965); Jennings

Randolph chaired the Senate Committee on Public Works. Understandably both men would be highly skeptical of any actions that would threaten either their constituents' interests or their state's most important industry. At the same time, the PHS also had to contend with Senator Muskie, chairman of the Public Works' Subcommittee on Air and Water Pollution who eventually gained a reputation as the Senate's most ardent environmentalist. Muskie's views ensured that he would not tolerate efforts to undermine the PHS's work.

The PHS's responsibilities and constituencies, as well as the lack of congressional guidance, seemed sure to generate acrimonious debate over the definition of clean air. The debate was not long in coming. Less than six months after the signing of the Clean Air Act, Muskie's subcommittee held five days of technical hearings in order to examine progress on pollution abatement. Several of the witnesses from industrial groups were not at all reluctant to challenge the need for any criteria document on sulfur oxides. One representative from the American Petroleum Institute suggested that the PHS was wasting its time studying sulfur oxides. After all, said P. N. Gammelgard, the public was not especially concerned over existing levels of sulfur oxides in the atmosphere, and even in cities with high concentrations of SO_2, no data existed to show that any detectable health effects occurred as a result.[21]

Gammelgard's arguments about the safety of SO_2 were echoed in nearly the same words by representatives of the coal industry when Muskie and his colleagues conducted additional hearings the following year. Such criticisms remained relatively mild, however, until the spring of 1965. At that time, no criteria had been published, but HEW was intent upon taking the lead in pollution abatement, and the best way to do so was by example. The Clean Air Act had not given the federal government any direct responsibility for air pollution activities in the states, so HEW could set an example only by instigating abatement actions at federal facilities.

HEW's choice of actions hit the oil and coal industries especially hard. HEW proposed that newly constructed federal facilities in Chicago, Philadelphia, and New York City be required to use coal containing 0.7 percent sulfur or less or residual oil containing 0.9 percent sulfur or less. For the largest federal facilities such as the Tennessee Valley Authority's power plants, HEW recommended that they burn the "lowest sulfur content fuel reasonably available."[22] In both instances, these emissions standards were apparently set so that the resulting air quality would not exceed what the PHS thought to be an acceptable level of SO_2. There was no question that the emission standards would reduce ambient levels of the pollutant, but the standards would be difficult to justify in the absence of a criteria document and adequate monitoring devices. For example, the PHS did not know that the coal regulations for the three cities would lead to acceptable levels of air quality because it did not know what volumes of SO_2 made the air unacceptable. Similarly its directive to burn fuel with the "lowest sulfur content reasonably available" was made without the PHS's demonstrating the existence of a pollution problem. If this reasonably available fuel was coal with a sulfur content of 1.2 percent, how did the PHS know that a coal with a sulfur content of 1.4 percent would not produce acceptable levels of air quality. Furthermore if low-sulfur fuels were expensive or more difficult to obtain than the fuels already in use, this might mean that the lower sulfur fuels were not "reasonably available."

The proposal created a storm of protest, and representatives of the National Coal Association and the American Petroleum Institute sought an immediate meeting with officials from the PHS and the Bureau of the Budget, which would have to approve the regulations. The PHS had estimated that the immediate effects of the regulation would be minimal because large amounts of low-sulfur coal were already available for purchase by the government. The oil and coal industries were not especially apprehensive about the immediate, limited effects that the PHS had projected; rather they feared that the proposals might be extended to all federal installations and eventually serve as a model for state and local agencies. In fact, the PHS did make the latter point.

The coal lobby was primarily interested in its own economic well-being, but the proposed rules had the potential to create far greater ramifications. Should state and local agencies adopt the guidelines, the result would be a massive change in patterns of fuel consumption. At the time of the proposal, no control technologies existed to remove sulfur oxides from stack gases after combustion but before entry to the ambient air. Some sulfur could be removed from coal before combustion, but this cleaning process could not be used on all coals and was not entirely satisfactory. In the absence of other means for reducing sulfurous emissions, the easiest way to do so is to limit the allowable sulfur content of the fuels burned or to switch to natural gas, which contains negligible amounts of sulfur. Thus if state and local governments followed the federal lead, states like Ohio, Illinois, Pennsylvania, and West Virginia, which produce large amounts of high-sulfur coal (coal with a sulfur content over 2 percent), would lose many traditional markets and thousands of jobs. Some supplies of low-sulfur coal do exist in these states, but many of these are high-grade metallurgical coals committed to the steel industry under long-term contracts. Accordingly strong incentives would exist to switch to available supplies of low-sulfur petroleum products or to natural gas.

Many oil companies were not enthused about the plan to increase consumption of low-sulfur petroleum products because it could upset profit and production schedules. For the most part, when power plants burn oil, they rely on residual fuel, which is the residue left over after the production of gasoline, jet fuel, kerosene, home-heating oil, and similar products. Without additional processing, the sulfur content of residual oil is always higher than the original crude oil's sulfur level since waste products such as sulfur accumulate in the residue. It is technologically possible to remove virtually all sulfur from residual oil, but it can be done only at a very high cost.

Within certain limits, refiners can alter their product mix to produce larger quantities of gasoline and relatively smaller quantities of other products and vice versa. In periods of low demand, refiners will seasonally reduce the output of certain products such as residual oil. In the face of a long-term absence of a market, however, refiners will attempt to reduce the production of low-demand products as much as possible in order to favor more profitable products such as gasoline. The latter situation occurred in the United States during most of the 1940s, 1950s, and even into the late 1960s. Few incentives existed to spend the extra money to lower the sulfur content of residual oil since the finished product would be more expensive per unit of electricity than would be coal or natural gas. In many instances, in fact, refiners had to sell their residual oils at less than cost.

In contrast to the disincentives associated with the production of residual oil in these years, the tremendous growth in the use of automobiles and jet planes required larger amounts of gasoline and jet fuel. These circumstances provided strong inducements for refiners to develop new technologies to maximize the production of high-value items like gasoline while reducing the production of low-value items such as residual oil. In this regard, American refiners were eminently successful. Beginning in 1943 and continuing for every year until 1966, the annual average yield of residual oil per barrel of crude oil from domestic refineries decreased. In 1943, the average yield of residual oil was nearly 30 percent of total output; by 1965, this percentage had declined to 8.1 percent.[23] Over the same period, there were sizable increases in the relative yields of gasoline and jet fuel. In short, few domestic refineries were equipped to respond to a rapid growth in demand for low-sulfur residual oil.

In addition to the complaints of industry, the Bureau of the Budget could not discount the probable economic dislocation that would result from reduced coal use. As J. Clarence Davies has noted, "A stringent sulfur standard threatened to strike a mortal blow to Lyndon Johnson's anti-poverty efforts in Appalachia, efforts to which he attached considerable importance." The budget office's concern was not matched by that of the PHS, which, says Davies, refused to discuss any economic problems.[24]

The arguments offered were sufficient to convince the Budget Bureau that the proposed sulfur regulations could not be justified. The budget office announced that the implementation of the regulation would be postponed pending answers to three questions. It wanted to know what levels of sulfur oxides are acceptable, what regulations would be needed to achieve such levels, and what sulfur levels of fuel oil the industry could provide with current technology and without disruption.[25]

The Budget Bureau's request for this information suggests that the PHS could not defend its proposal adequately. The bureau's questions, though simple, struck at the core of the debate. The budget office could hardly approve such a controversial proposal if the PHS could not indicate what levels of sulfur oxides are harmful. Of equal interest is the fact that the Budget Bureau introduced economic considerations into its decision to postpone the regulations. The PHS had considered health to be the primary determinant of its sulfur regulations and was not prepared to include economic variables in its calculations. The budget people disagreed with this exclusion; the need for improved levels of air quality would have to be balanced against economic and political considerations. Until the PHS incorporated these considerations, the Budget Bureau would not consider the proposal.

The decision to postpone the regulations did not end the PHS's embarrassment at having its proposal soundly rejected. Shortly after the decision, the House Commerce Committee's Subcommittee on Public Health and Welfare conducted hearings on a series of plans to regulate the emission of pollutants from automobiles. Despite the purpose of the hearings and the fact that the proposal had already been withdrawn, much of the discussion focused on the PHS proposal. The most vehement critic was Joseph E. Moody, president of the National Coal Policy Conference (NCPC). Claiming that the proposal would impose unnecessarily stringent restrictions on the use of coal, he asked: "What is an acceptable level of sulfur dioxide emissions consistent with the Nation's well-being and health and national economy. . . . As far as I can determine, this has never been determined."[26]

The coal industry was concerned about the need for a reduction in sulfur emissions, but Moody saved most of his criticisms for the possible economic impact of the proposal. He claimed that if state and local agencies adopted regulations similar to the federal ones, the effect on his industry would be devastating. Referring to a PHS study, Moody noted that the average sulfur content of coal burned by utilities in 1963 exceeded 2.5 percent; less than 5 percent of all the coal burned in that year could meet the proposed level of 0.7 percent sulfur.[27] The oil industry was similarly apprehensive since the sulfur content of residual oil that utilities used ranged from 1.5 to 2 percent.

The Coal Policy Conference made two suggestions to members of the House subcommittee. First, if the government was going to impose sulfur limitations, it should also develop economic means to remove sulfur from fuels either before or after they are burned. In effect, the conference advocated that improvement in the nation's health should be tied to the government's willingness to ensure the coal industry against any adverse economic effects. In Randall Ripley's terminology, the coal interests were pressing for regulation that was "either favorable or at least not too bothersome to their interests."[28] Second, the coal lobby suggested that future decisions affecting energy and the national economy be handled by Congress "rather than by a segment of the executive branch whose field of responsibility is limited and narrow in scope."[29] This second argument can be seen as another typical response to regulatory policy. Dissatisfaction with a regulatory agency's performance often leads to an appeal to the entire Congress.

The opposition of the fuel industries and the Bureau of the Budget was persuasive. Congressman Oren Harris, chairman of the House Commerce Committee in 1965, was puzzled by HEW's action and said he could not understand why the proposal had been made. He asked and received assurances from HEW that it would withdraw the proposals temporarily for "further consideration and study."[30]

This further consideration took place in the fall of 1965, and by early 1966, HEW and the PHS were ready to try again. This time, however, the Budget Bureau made some effort to confer with the oil and coal interests before issuing any proposals. At the conclusion of the consultations, President Johnson signed an executive order that outlined the government's objectives for clean air at its own facilities. The order declared that the government's policy would be to prohibit emissions of most pollutants from federal facilities "if such emissions endanger health or welfare"; emissions that might be injurious or hazardous to people, animals, vegetation, or property" would be minimized. Even within the order, however, contradiction was evident because the emissions of sulfur oxides at federal sites would only have to be "minimized to the extent practicable," regardless of their impact.[31] For sulfur oxides, therefore, their reduction would be related to such considerations as the coal industry's economic well-being and not to health considerations, as Johnson's order had originally stated.

A few days after the president's action in May 1966, the secretary of HEW issued specific guidelines for implementing the executive order. The new HEW rules required that both new and existing federal facilities "burn the lowest sulfur content fuel that is reasonably available." Reasonable availability would be determined by such factors as "price, firmness of supply, extent of existing pollution, and assurance of supply under adverse weather and natural disaster conditions." The regulations also required the secretary of HEW to set specific sufur-in-fuel limits for federal facilities in the nation's largest and most densely populated cities within six months of the regulation's effective date.[32] Unlike the proposed regulations of the previous year, HEW indicated that it would not set any limits on a fuel's use until after it had consulted with affected parties and other government agencies.

The coal and oil industries knew about the guidelines in advance but were unprepared for HEW's specific recommendations. In late 1966, still without criteria for sulfur oxides, HEW issued regulations for federal facilities in Chicago, Philadelphia, and New York City. To achieve a desired level of 0.015 parts per million of SO_2 on an annual basis, HEW said that after October 1, 1968, the sulfur content of coal burned in New York City should not exceed 0.2 percent and that of oil should not exceed 0.3 percent. For Chicago and Philadelphia, 0.4 percent coal and 0.6 percent oil would be required. As the figures in table 3–1 show, these new limits were considerably more stringent than those that HEW had proposed a year earlier.

In accordance with its earlier guidelines, HEW allowed a short period for public comment. As might be expected, the coal interests were unhappy. Although it was primarily concerned with the economic consequences of the sulfur limits, the industry did not abate its criticisms of studies linking sulfur oxides to

Table 3–1
HEW's Recommended Sulfur-in-Fuel Limits for Federal Facilities
(percentage)

	1965	*1966*	*Change*
Chicago and Philadelphia			
Coal	0.7	0.4	−57
Oil	0.9	0.6	−33
New York City			
Coal	0.7	0.2	−29
Oil	0.9	0.3	−67

ill health. Indeed it availed itself of many opportunities to testify before congressional committees about the questionable need for criteria. In testimony before the House Subcommittee on Science, Research, and Development in mid-1966, for example, a spokesman from the National Coal Association argued that a causal relation between sulfur oxides and ill health had yet to be established. The lack of such data, said one coal lobbyist, was due to the inability to assess the damages associated with sulfur oxides: "One cannot see them or feel them . . . and the concentrations of sulfur oxides in the air are for the most part so low that we are not aware they exist."[33]

This statement seems to have been a naive attempt to divert attention away from SO_2. The issue was not whether the average citizen can see or detect the pollutant but the pollutant's effects. To discourage any notions that the coal industry was unconcerned about these effects, it did call for further research so that criteria could be determined on the basis of "factual information and not on emotions." Repeating previous arguments, the coal association urged that the availability of control technology should determine the imposition of acceptable levels of sulfur oxides "until the *exact* levels of pollution which are dangerous to man have been established."[34] In other words, the industry intended that acceptable levels of air quality should not be set in the absence of incontrovertible data but rather on the basis of polluters' willingness to develop and utilize new control technologies or industrial processes. In this way regulation would be acceptable to the regulated interests.

Publishing the Criteria

These considerations did not deter further work on the criteria. By early 1967, the PHS's initial criteria document on sulfur oxides was nearing completion. Before final publication, however, the health service wanted to circulate the criteria for a final appraisal. At this point, the fuel industries intensified their efforts either to confuse the issues or to delay publication altogether. Joseph Moody of the NCPC met with officials of HEW several times in January and February 1967 and continually challenged both the need for a criteria document and the data supporting the findings.

The coal industry's major congressional ally also attempted to influence HEW's forthcoming criteria. Senator Randolph wrote to the head of HEW's National Center for Air Pollution Control in mid-January and asked that "regulations and criteria [for sulfur oxides] . . . not be issued on a basis of outpacing technology and supplies." Randolph further requested "complete and irrefutable proof of necessity" for the forthcoming criteria.[35] Randolph followed this letter with another to President Johnson a few days later. He repeated his concerns and added that policy makers were placing undue emphasis on the abatement of sulfur oxides.[36] As Ralph Nader's report on air pollution later noted, Randolph's letters to the executive branch "could not have made the coal interests happier if they had been drafted by Moody himself," which was probably the case.[37] Randolph had not made any specific suggestions about how the criteria could be improved; he was interested only in delay. Moreover his requests were implausible. After all, what is "complete and irrefutable proof" and how does one know when it exists? As chairman of the Senate committee with jurisdiction over air quality legislation, Randolph would ordinarily be expected to support HEW's efforts to abate pollution. The Clean Air Act had cleared his committee, and the PHS was only trying to implement what it believed to be congressional intent. That Randolph was openly nonsupportive did not bode well for the first criteria.

These pressures did not change HEW's intentions. On March 22, 1967, the secretary of HEW gave final approval to the sulfur regulations applicable to federal facilities. More important, after more than three years' effort, the manual, *Air Quality Criteria for Sulfur Oxides,* was issued on the same date. These criteria, which were the first issued for any pollutant, were said to "summarize what is known about the effects of pollutants in the atmosphere to provide a realistic base for selecting air quality standards."[38] The new criteria reported on several hundred studies, but many of these had been done in the 1940s and 1950s; reference was even made to one published in 1907. Fewer than five studies had been completed in 1966. Thus it was not surprising that the criteria pointed out that information on the effects of sulfur oxides was limited. The report also added an important note of caution:

Although the criteria presented . . . index the effects of the oxides of sulfur, these effects do not necessarily, nor in fact actually, derive solely from the presence of sulfur oxides in the atmosphere. They are for the most part the effects observed when various concentrations of sulfur oxides, along with other pollutants have been present in the atmosphere.[39]

In making this statement, the report explicitly recognized that SO_2 by itself might not be harmful at levels found in the environment. Although this meant that other pollutants, such as sulfates or sulfuric acid, might be the primary or determinant causes of adverse effects, the findings were expressed in terms of concentrations of SO_2 because only SO_2 "has been measured in most reported investigations of sulfur pollution."[40]

Figure 3-1 represents a diagram that HEW used to summarize the relation between levels of SO_2 and times of exposure. On the basis of previous studies, HEW recommended that acceptable levels of SO_2 could be defined as those that fell below and to the left of the cross-hatched area in the figure. Numerically this means that levels of SO_2 should not exceed annual average concentrations of 0.015 ppm. Above this level, the document stated, impairment to the health of sensitive or susceptible groups of people was likely. This recommendation was very stringent because levels of SO_2 often exceeded 0.170 ppm in some American cities in the mid-1960s. Additionally many other cities had levels below 0.170 ppm but still far above what HEW now considered to be acceptable. For periods of less than a year, HEW recommended the levels listed in table 3-2.

In accordance with the Clean Air Act, the states and local governments could use the criteria and their recommendations to set air quality standards. In order to meet these standards, limitations on emissions of SO_2 would have to be established based on such considerations as the relative cost and technical feasibility of pollution abatement measures. Hence eventual attainment of the standards would hinge on the availability of control technology, as well as polluters' willingness to accept the costs of the technology.

The criteria assured that coal and oil would be the two industries most likely to face decisions on cost and feasibility. According to the criteria, over 23 million tons of SO_2 had been emitted into the air over the United States in 1963. Of this amount, approximately 60 percent resulted from the combustion of coal and slightly over 22 percent came from refinery operations or the combustion of residual oils.

Conclusions

Many people believe that crises or disasters provoke immediate responses from governments concerned for their citizens' welfare. While this is true in many instances, the nineteen-year gap between the Donora incident and publication of the SO_2 criteria suggests that rapid response is not typical for equally important but more subtle problems like air pollution. In the years after the Donora episode, the federal government seemed to express a continued reluctance to become involved in solving the growing problems of air pollution. The 1955 legislation did little more than acknowledge that pollution was a problem. The Clean Air Act of 1963 repeated this acknolwedgment and encouraged but did not require the states to address the problem.

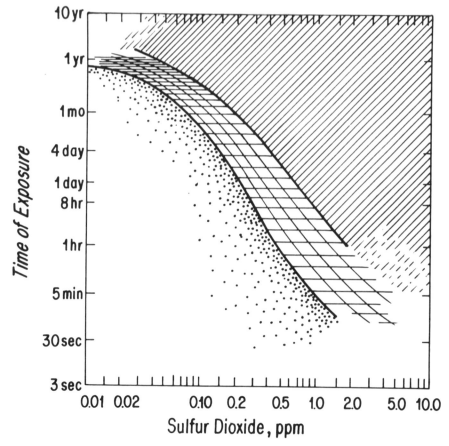

Range of concentrations and exposure times in which deaths have been reported in excess of normal expectation.

Range of concentrations and exposure times in which significant health effects have been reported.

Ranges of concentrations and exposure times in which health effects are suspected.

Source: HEW, PHS, *Air Quality Criteria for Sulfur Oxides* (Washington, D.C., 1967), p. liii.

Figure 3-1. Effects of Sulfur Oxides Pollution on Health

At best, federal initiatives between 1963 and 1967 were haphazard. Inadequate scientific and political support marked HEW's early proposals for controlling sulfur oxides. There is no doubt that HEW had to temper its initiatives because of the Bureau of the Budget's insistence that economic factors be considered, President Johnson's ambivalence about the control of SO_2, and the

Table 3-2
HEW's Recommendations for Acceptable Levels of Air Quality

Time Period	Maximum[a] (ppm)	One Percentile[b] (ppm)
Twenty-four-hour average	0.05-0.08	0.04-0.06
One-hour average	0.12-0.20	0.05-0.11
Five-minute average	0.10-0.50	0.05-0.14

Source: HEW, PHS, *Air Quality Criteria for Sulfur Oxides* (Washington, D.C., 1967), p. lvi.
[a]Maximum: the highest value expected to be reached throughout the time period.
[b]Percentile: the percentage of time a specified value is allowed to be equaled or exceeded during the time period. Thus, one-percentile values could be reached or exceeded 1 percent of twenty-four hours, one hour, or five minutes.

fuel industries' unwillingness to tolerate limited but potentially disruptive restrictions on fuel use at federal facilities. Thus the persistent lack of support for HEW's activities was not at all predictive of the criteria document that finally emerged in 1967. The document placed the public's health above all other considerations. Political opposition and gaps in scientific understanding would no longer justify delay, or at least this was the position of HEW. The sole emphasis on public health also meant, however, that HEW and the PHS had neglected the bargaining and negotiation with affected interests that typically precedes implementation of regulatory statutes.

Notes

1. Cited by Charles O. Jones, *Clean Air* (Pittsburgh: University of Pittsburgh Press, 1975), p. 26.
2. PL84-159, sec. 2(b), July 14, 1955.
3. Ibid., sec. 3, 4.
4. Jones, *Clean Air,* p. 35.
5. Randall B. Ripley, "Congress and Clean Air: The Issue of Enforcement, 1963," in F.N. Cleaveland, ed., *Congress and Urban Problems* (Washington, D.C.: The Brookings Institution, 1969), p. 235.
6. Clean Air Act, PL 88-206, sec. 1 (a)(3), December 17, 1963.
7. HR 4415, 88th Cong., lst sess., 1963.
8. U.S. Senate, Committee on Public Works, Special Subcommittee on Air and Water Pollution, hearings, *Air Pollution Control*, 88th Cong., 1st sess., 1963, p. 283.
9. S1040, 88th Cong., 1st sess., 1963.
10. Senate, *Air Pollution Control,* p. 86. Emphasis added.
11. Clean Air Act, sec. 3(c)(2).

12. J. Clarence Davies III, *The Politics of Pollution* (New York: Pegasus, 1970), pp. 158–159.

13. Clean Air Act, sec. 5(g).

14. Ibid., sec. 1(b)(1).

15. Murray Edelman, *The Symbolic Uses of Politics* (Urbana: University of Illinois Press, 1964), chap. 1, 2.

16. Ibid., p. 141.

17. Theodore J. Lowi, *The End of Liberalism* (New York: W. W. Norton, 1969), p. 299.

18. U.S. Senate, Committee on Public Works, Report No. 88-638, *Clean Air Act*, 88th Cong., 1st sess., 1963, p. 7.

19. HEW's decision to focus initial attention on sulfur oxides was subsequently reinforced in late 1965 with the publication of a report by the President's Science Advisory Committee, which listed sulfur oxides first among a group of pollutants with the highest priority for additional study. See "Restoring the Quality of Our Environment" (1965).

20. John C. Esposito, *Vanishing Air* (New York: Grossman, 1970), p. 103.

21. U.S. Senate, Committee on Public Works, Special Subcommittee on Air and Water Pollution, hearings, *Clean Air, Part 2,* 88th Cong., 2d sess., 1964, p. 1326.

22. *Wall Street Journal*, June 11, 1965, p. 1.

23. American Petroleum Institute, *Petroleum Facts and Figures*, 1971 ed. (Washington, D.C.: American Petroleum Institute), p. 203.

24. J. Clarence Davies to the author, July 1978.

25. Cited by *Journal of the Air Pollution Control Association* 15 (August 1965): 380.

26. U.S. House of Representatives, Committee on Interstate and Foreign Commerce, Subcommittee on Public Health and Welfare, hearings, *Clean Air Act Amendments,* 89th Cong., 1st sess., 1965, p. 337. The Bureau of the Mines, an ally of the coal industry, supported Moody's argument. The bureau said that the PHS requirements were unrealistic in the absence of data justifying them. See *Journal of the Air Pollution Control Association* 15 (July 1965): 332. Interestingly the bureau was not at all reluctant to criticize the health basis of PHS's recommendation in 1965, but the following year the bureau's director told a congressional subcommittee that his agency was not competent to discuss the health effects of SO_2. See U.S. House of Representatives, Committee on Science and Astronautics, Subcommittee on Science, Research, and Development, hearings, *The Adequacy of Technology for Pollution Abatement*, 89th Cong., 2d sess., 1966, 1:269.

27. House, *Clean Air Act Amendments*, p. 343.

28. Randall B. Ripley and Grace A. Franklin, *Congress, the Bureaucracy, and Public Policy* (Homewoood, Ill.: Dorsey Press, 1976), p. 97.

29. House, *Clean Air Act Amendments,* p. 344.

30. U.S. House of Representatives, Committee on Interstate and Foreign Commerce, Report No. 899, *Clean Air and Solid Waste Disposal Acts*, 89th Cong., 1st sess., 1965, p. 3.

31. Executive Order 11282, May 28, 1966.

32. *Federal Register* 31 (June 2, 1966): 7903-7904.

33. House, *Adequacy of Technology,* 2:909.

34. Ibid., p. 910. Emphasis added.

35. Cited by Esposito, *Vanishing Air*, p. 281. Randolph's letter was dated January 12, 1967.

36. The letter, dated January 16, 1967, is reprinted in U.S. Senate, Committee on Public Works, Subcommittee on Air and Water Pollution, hearings, *Air Pollution—1967 (Air Quality Act), Part 4*, 90th Cong., 1st sess., 1967, pp. 2455-2456.

37. Esposito, *Vanishing Air*, p. 281. In a letter to the author, J. Clarence Davies suggested the origin of Randolph's comments.

38. HEW, PHS, *Air Quality Criteria for Sulfur Oxides* (1967), p. iv.

39. Ibid., p. v.

40. Ibid, p. li.

4 Reaction and Response

The publication of the criteria on sulfur oxides evoked an immediate and bitter response from the coal, oil, and electric utility industries. These groups realized that they could not oppose clean air, but they also knew that they had to refute the PHS's data on sulfur oxides. Left unchallenged, these data could lead to major changes in fuel use, impose important restrictions on the industries' operations, and perhaps even jeopardize the existence of the coal industry. Fortunately for these affected interests, their attack was launched at the same time that the Congress was beginning its consideration of revisions to the Clean Air Act, enabling the industries to vent their displeasure not only in their own trade publications but before several congressional committees as well. If the industries could convince these committees that the air quality criteria recommended unnecessarily stringent levels or that they would create economic turmoil, then the industries would find themselves in a position to undermine the PHS's efforts. Alternatively if the interests presented a weak or unpersuasive case, the prospects for further restrictions would be enhanced.

In one of his first messages to the Ninetieth Congress, President Johnson made clear his preference for the latter course of action. His speech, "Protecting Our National Heritage," focused much attention on the problems associated with air pollution and especially with sulfur oxides, which he alleged, "threaten the environment of almost every city and town in America."[1] To combat the growing problem of air pollution, Johnson believed that comprehensive changes were in order. At the time of his speech (January 30, 1967), the PHS had not issued a criteria document, but this did not prevent Johnson from recommending national emission standards for major sources of air pollution. Such standards, said Johnson, "will establish pollution limits that a given industrial plant may not exceed.... Our aim is to provide uniformity and stability in pollution control levels in cooperation with industry and local governments."[2]

The Clean Air Act had required the national government only to issue air quality criteria that state and local agencies could, if they wished, use to set air quality and emission standards. When Johnson made his recommendations, he knew that the document on sulfur oxides would soon be available, but he also understood that publication would be only the first step in a long process. Under the provisions of the 1963 law, many years would pass before all states adopted emission standards and procedures to implement these standards. Johnson's impatience with this process was reflected in his proposal that the Air Quality Act of 1967 eliminate some state and local discretion by requiring the secretary of HEW to publish industry-wide emission standards. The directives would be based on air quality criteria developed as a result of the earlier legislation.

The coal and oil industries were understandably skeptical about such an approach. HEW was already known to favor strict limitations on the allowable sulfur levels of coal and oil, and the fuel industries did not want to see such curbs arbitrarily extended to all states. The industries preferred that state and local agencies retain their freedom to impose or reject any sulfur regulations. The industries believed that major coal-producing states were unlikely to impose too stringent regulations. For many other states, the coal and oil industries could rely on the effectiveness of their own lobbying, the absence of effective anti-pollution agencies among most states, and a widespread unwillingness to impose restrictions without a compelling need to do so.

The fuel industries adopted a twofold strategy in their efforts to fight federal encroachment. First, they decided to attack the PHS's credibility and scientific objectivity. The goal would be to convince policy makers that sulfur oxides are not injurious to health and to claim otherwise would be a serious misrepresentation of scientific knowledge. At the least, these attacks might compel further study and, consequently, delay the imposition of air quality or emission standards. Second, the industries would portray the economic havoc that would result if national emission standards were adopted or if state and local governments translated the new sulfur oxides criteria into air quality or emission standards.

Because the PHS had submitted a preliminary draft of the criteria to the National Coal Association with a request for comments, the coal interests were prepared to respond to the criteria document when it was finally published. The coal association had hired Hazleton Laboratories, a private scientific concern, to prepare a critique of the document.[3] Not unexpectedly, the assessment was both thorough and highly abrasive. It claimed several major weaknesses in HEW-PHS's efforts. First, its author maintained that the criteria were grossly inadequate and had ignored much relevant information. To support this claim, one study was cited that had listed over 2,200 references to sulfur oxides; the criteria listed approximately 340, and many of these were supposedly irrelevant and represented "erudite padding."

Second, the criteria were said to be suspect because many of the data were not applicable to day-to-day situations with sulfur oxides. As an example, the report noted that HEW had assumed, incorrectly, that results from animal studies are directly applicable to humans.

Third, HEW allegedly gave "unwarranted and undue weight to a limited number of epidemiological surveys whose data are inconclusive [and] open to several interpretations." The Hazleton report also expressed concern that HEW had eliminated the scientific qualifiers found in many of the original studies and had treated all studies with equal weight and without regard for their relative merits. A review of the criteria's supporting data certainly lent some credence to this argument. The PHS had constructed a summary diagram on the effects of sulfur oxides (see figure 3-1) on the basis of studies done in at least a half-dozen countries, with six different methods of measuring sulfur dioxide, and with studies conducted over thirty years. One study had not used measuring devices; levels of SO_2 had simply been estimated several years after an air pollution episode. The diagram had also combined epidemiological with animal and clinical studies. Regardless of the methods of the various studies, they had been given equal weight in construction of the diagram.

Finally the critique maintained that the criteria document was replete with bias, oversimplification, prejudiced conclusions, unscientifically doctrinaire viewpoints, and "bold, unqualified numerical certainties."[4] The report concluded that HEW and the PHS had done a disservice to science by misrepresenting a relationship between sulfur oxides and harm to the public's health and welfare when, in fact, such a relationship had not been established.

Some support for Hazleton's claims came from two unexpected sources. Shortly after the criteria were published, the *Air and Water News*, a weekly environmental newsletter, reported that several environmental specialists who had reviewed the criteria felt that the scientific justifications were "frequently vague, incomplete or quoted out of context." Furthermore these reviewers expressed considerable skepticism about the research work used to justify the claim that extremely low levels of sulfur oxides can cause harm.[5] A far more damaging report was published in the *Environmental Health Letter*. According to this publication, the President's Science Advisory Committee had convened a group of scientists to evaluate the criteria. The group reportedly "expressed almost unanimous amazement at the inadequacy of the scientific data on which the criteria were based."[6] Although the PHS could question the motives of the fuel industries' criticisms, the two independent appraisals could not be brushed aside as easily. The health service had said the criteria would be noncontroversial, but the extent and the breadth of the criticisms showed that this would not be the case.

In addition to criticizing the criteria on scientific grounds, the coal and oil interests also expressed concern because of the possible economic consequences of restrictions on sulfur-bearing fuels. Many oil companies believed that a greater demand for low-sulfur oil might jeopardize their existing investments in high-sulfur fields, and the coal interests were apprehensive that the PHS's recommendations would encourage electric utilities to curtail their use of coal. The industries were surely anxious about their own economic self-interests, but they were careful to couch their complaints in terms of the economic damage that would occur to the nation. From this perspective, the definition of clean air would include not only a scientific collection of existing data but also an extended discussion of Americans' economic well-being and life-style. One coal executive maintained that if the recommendations were put into effect, Americans "would be faced with a blackout which would blanket substantially all the Nation and reduce the wheels of industry to a crawl for years and years."[7]

The success and impact of these attacks can best be gauged by their reception in the White House and the Congress. At the White House, the pleas for relief did not go unheeded. President Johnson intervened on behalf of the coal industry by submitting a request to the Congress for a supplemental appropriation of $2.7 million for research on controlling pollution from sulfur oxides.[8] Johnson's request more than doubled HEW's budget allocated to the development of control technology. In addition to providing a sympathetic response to the coal industry's plight, Johnson's action gave support to the argument that the government had a responsibility for ensuring that potential restrictions on coal did not exceed the availability of control technology. This precedent, though not extended to other industries such as steel or automobiles, allowed the fuel industries to claim that what they or the electric utilities could not do to control pollution was the government's responsibility.

The decision to boost HEW's research budget virtually exonerated the industries from the job of developing their own control equipment. Realistically modest research expenditures on their part would serve as an indication of good-faith efforts. Once these efforts had been made, they could sit on the sidelines until the government developed control technologies acceptable to the industries. The president's position on this issue meant, of course, that any discussion of national emission standards for coal- or oil-burning electric utilities would be futile in the absence of governmental progress on pollution control devices. This approach would certainly be agreeable to the fuel industries, but they realized that congressional assent for the strategy would be necessary.

Action in the Congress

After President Johnson's decision to boost research expenditures, attention turned to Senator Muskie's Subcommittee on Air and Water Pollution, then in the midst of hearings on new air quality legislation. Much of the initial testimony focused on the president's proposal for national emission standards. Spokesmen for the Johnson administration argued that national standards were necessary. Past experience, said HEW Secretary John Gardner, showed that state and local governments would not establish control measures beyond those set up elsewhere for fear of finding themselves at a competitive disadvantage. Supposedly states with strict environmental standards discouraged new industrial development within their boundaries while encouraging it in states with less stringent regulations. National standards were intended to avoid this problem.

A further purpose of national emission standards would be to stimulate technological development. An important theme underlying the Clean Air Act was voluntary abatement by polluters, but by 1967 it was obvious that anti-pollution agencies could not count on such volunteerism. Some mechanism was needed to force the installation of control devices. Mandatory emission standards would accomplish this task, claimed Gardner, and would "undoubtedly provide an attractive economic incentive to the development of control technology and will result in better, cheaper and, more important, widely applicable ways of reducing pollution from specific sources."[9]

Apparently the theme was that national standards would force the development of control technology. This was not, however, the administration's intent. Instead Gardner explained that federal emission standards would be established only after "appropriate consideration" had been given to the standard's economic and technical feasibility. The standards would be tightened as technology improved. Accordingly acceptable levels of air quality would not be achieved if control technologies were either unavailable or too expensive for polluters. This approach had at least two major weaknesses. First, it would allow polluters to define economic and technical feasibility. In the absence of sufficient research and development expenditures, polluters could rightly claim that better control devices were either unavailable or, at best, "just around the corner." In terms of the coal and electric utility industries, President Johnson's earlier funding decision meant that acceptable levels of SO_2 would depend on the federal government's ability to develop appropriate control technology and convince the industries that this nonproductive technology was economically attractive. The former task could be costly and time-consuming, the latter task nearly impossible.

Second, the approach could diminish the importance of the public's health as a concern in setting emission standards. As John T. Middleton, director of HEW's newly organized National Center for Air Pollution Control, noted, consideration of health effects might warrant more stringent standards, but they would not be imposed in the absence of economically and technically feasible control devices.[10]

These kinds of problems raised understandable skepticism among many senators who wanted to know which industries would feel the initial impact of national emission standards. HEW and the PHS offered contradictory testimony. One of HEW's undersecretaries indicated that power plants would be obvious candidates for such standards but that "the present state of the art in terms of sulfur emissions would not permit an early application of that standard nationally."[11] An official from the PHS testified that power plants would be among the first to be regulated.[12]

The administration's inconsistency on the likelihood of national standards for power plants turned out to be meaningless in view of its inability to present a convincing case in their favor. In addition, Senator Muskie's well-known opposition to them compounded the administration's difficulties. In the end, the Senate firmly rejected the idea of national standards but accepted the idea that the availability of control technology should precede regulation. Consequently the Senate intended that control equipment should be installed when it became available; this position paralleled the coal industry's belief that regulation should not outpace technology. Unlike the coal industry, however, Muskie believed "that the technology of pollution control has progressed sufficiently to achieve reasonable standards of air quality."[13]

Although Muskie was opposed to federally mandated emission standards, he was in favor of national air quality criteria that could be used to set emission standards on a regional basis. Thus much of the subcommittee's attention turned to a discussion of air quality criteria after publication of the document on sulfur oxides. Muskie could not be displeased that criteria had been issued, but he was concerned that only one document had been produced after nearly four years of effort. Unlike Muskie, Senator Randolph's interest was in the content of the first document, not the pace of publication. His earlier letters to HEW's air pollution unit and to President Johnson revealed his apprehension, and this was frequently reflected in his comments to witnesses testifying before the committee, about the criteria. At one point, Randolph even suggested to the surgeon general that confusion surrounded his agency's work on the criteria.[14] The surgeon general tried to be diplomatic in his response but still felt compelled to rebut Randolph's concern. The PHS had concluded that no revisions in the criteria were necessary. Accordingly, it believed that state and local governments could establish air quality standards on the basis of scientifically valid criteria.

The Senate was less confident that the PHS about the validity of the scientific data relating SO_2 to ill health. The Senate's report on the Air Quality Act of 1967 noted that the controversy over the criteria compelled further research conducted with "diligence and perseverance."[15] In fact, the Senate accepted, without change, several recommendations proposed by Hazleton Laboratories, which the National Coal Association had hired to review the criteria on sulfur oxides. These recommendations were so extensive as to suggest that little previous research had ever been done on sulfur oxides.

This impression was false, but the Senate was evaluating the criteria not only on their scientific merits but also on their economic and political acceptability to affected interests. While the criteria on sulfur oxides did have problems, they were at least a reasonable effort in view of the lack of precise legislative guidelines and the Congress's own confusion over criteria and standards. The PHS had tried to compile an intelligible summary of hundreds of studies, and the criteria represented a concise, but comprehensive analysis of what was then known about sulfur oxides and adverse effects. The Clean Air Act had not required the PHS to consider the social or economic ramifications of its analysis, and even if it had wanted to, the agency had neither the resources nor the expertise to study these issues.

This remoteness from political concerns worked to the disadvantage of the PHS. While recognizing that criteria documents would be the principal basis for any progress toward cleaner air, at Senator Randolph's suggestion the Senate required that all criteria previously issued be reviewed and reevaluated in accordance with new guidelines and, if necessary, modified and reissued.[16] The Senate further suggested that the PHS not await final congressional action before initiating this review. The message could not have been clearer. The Senate's report on the Air Quality Act told the PHS that even if the House of Representatives did not accept the Senate's version of the new law, the PHS would still be expected to redraft the criteria, or at least to provide more palatable conclusions.

The Senate's reaction was a devastating blow to the PHS. Only one set of criteria had been issued, and the service had staked its professional reputation on the document's accuracy and validity. The PHS argued that of all the evidence on air pollution, the data were most compelling in the case of sulfur oxides. In his testimony before Muskie's subcommittee, for example, the surgeon general, William Stewart, had emphasized that "the evidence on sulfur dioxide is sufficient that one would be negligent in saying that this does not have an effect on health and not do something about it."[17] The Senate disagreed, rejected the position of the PHS, and said that any action based on the initial document would be premature.

The Senate's harsh reaction was not unexpected. Politicians tend to accept scientific advice when it promotes, coincides, or does not interfere with the interests of private economic groups.[18] In contrast, scientists find their influence diminished when their data are inconclusive, subject to competing explanations, or likely to lead to or support regulation. These traits characterized the criteria and offer a further explanation for the Senate's action. In sum, the Senate wanted more compelling evidence than the PHS had provided. Moreover by requiring this evidence, the Senate could avoid further controversy with the fuel interests, which had found the criteria to be anathema. Often when a recommendation is controversial, the solution is to require further study. This was the Senate's choice.

This decision put the Senate in the peculiar position of favoring both speed and time-consuming deliberateness. While insisting on speed, the Senate added provisions that would delay. The Senate report stressed that the nation's air pollution program could not be implemented successfully unless the publication of criteria was accelerated.[19] In contrast, the Senate also proposed two procedures that would hinder further publications. Again in response to Senator Randolph, the Senate first proposed a new requirement that no further documents be issued until HEW had consulted "with appropriate advisory committees and Federal departments and agencies."[20] Randolph's original idea had been to create an interdepartmental committee of cabinet officers, as well as separate advisory committees, for each major industry with members selected "from among the persons engaged in such industry who are knowledgeable concerning air quality, or the economics or technology of the industry."[21] Other than the secretary of HEW (or his representative), who would chair each committee, there was no provision for public or nonindustrial participation.

Under Randolph's plan, the industrial committees' main responsibility would have been to evaluate and comment on all criteria and standards to ensure consistency with the purposes of the legislation and "the maximum utilization of the Nation's industrial resources." As a result, Randolph's amendment would have had the effect of allowing an industry group to veto or modify substantially scientific findings (in the case of criteria) on nonscientific grounds. The Clean Air Act had not required formal consultation with any groups outside of HEW, and Randolph's desire for change was an unequivocal response to charges that HEW and the PHS had ignored the concerns of the coal and oil industries.

The Senate did not accept Randolph's idea without change. Instead it proposed the creation of an Air Quality Advisory Board, with members appointed by the president, and other technical committees, with members appointed by the secretary of HEW. These committees would be composed of environmental officials, medical and scientific personnel, and industry experts in pollution abatement. One of their major responsibilities would be to assist HEW and PHS in the development and implementation of air quality criteria and standards. The Senate plan would open decision making to a larger number of interests than Randolph had intended, but it also meant that the PHS could not complete its obligatory review of the criteria on sulfur oxides until after members of the various committees had been selected and consulted.

The second way in which the Senate ensured delay in the publication of revised criteria was in its requirement that HEW compile and publish a document discussing recommended control techniques. After consultation with other federal agencies and the newly formed advisory committees, HEW was supposed to issue a report outlining "current technology and the economic feasibility of achieving various levels of air quality as specified in the criteria including alternative control techniques and their economic feasibility."[22] The technology report would be published at the same time as the criteria. In requiring simultaneous publication, the Senate was insisting that the quest for cleaner air not outpace the availability of control equipment.

Although the Senate responded unfavorably to the initial criteria, it did not reject their importance as precursors of regulation. President Johnson's request for national emission standards was relegated to a two-year study, but in their place the Senate proposed that states be required to develop their own air quality and emission standards. The Clean Air Act had directed that criteria be published for informational purposes only, but the Senate's stance in 1967 meant that all new criteria would require the states to set their own air quality standards. This mandatory action would proceed as follows: after publication of criteria and recommended control techniques for each pollutant, states would have ninety days to file a letter of intent with HEW acknowledging that they would establish air quality standards for the pollutant within six months. These standards could be "influenced not only by a concern for the protection of health or welfare, but also by economic, social, and technological considerations."[23] The standards would be applicable to air quality control regions, which HEW would establish within eighteen months of the law's passage on the basis of such considerations as atmospheric conditions, jurisdictional boundaries, and levels of industrial activity. Each state would also develop an implementation plan that would outline how it intended to achieve the air quality standards. The plans could consider the economic and technical feasibility of compliance.

In its version of the Air Quality Act, the Senate left unchanged the ambiguity that existed in the Clean Air Act over the definition and use of criteria. Criteria are supposed to be descriptive summaries of scientific knowledge, but in the revised law the Senate retained a prescriptive connotation: the secretary of HEW was directed to develop and issue "such criteria of air quality as in his judgment may be requisite for the protection of the public health and welfare."[24] This would require a judgment of acceptability and not merely a scientific assessment of risk. Similarly the Senate said that the secretary could not approve the states' air quality standards unless they were consistent with the air quality criteria and recommended control techniques. This consistency requirement created still more ambiguity. Did *consistent* mean *equal to, compatible,* or *congruous,* or was it simply a politically acceptable phrase that assured the need for negotiation with affected industries? Did the states' air quality standards have to be consistent with the criteria's scientific findings or with HEW's judgment of acceptability? If the former, then the Senate was merely insisting that the criteria be a source of information for state action. If the latter, it meant that HEW would actually be setting minimum national air quality standards, which the Senate had already rejected.

Once the Senate finished its deliberation on the Air Quality Act of 1967, it was another month before work began in the House Committee on Interstate and Foreign Commerce. With few members of the House Committee strongly interested in the problems of air pollution, the hearings were much less comprehensive than they had been in the Senate. Although the members of the Commerce Committee spent only eight days on hearings, this did not mean that HEW was viewed any differently than in the Senate. As an example, Congressman John Dingell of Michigan expressed concern about the delay in publishing criteria, and he accused the government's air pollution officials of an unwillingness to grapple with the problem and of a dereliction of their responsibilities.[25]

After Dingell's tongue lashing, officials from the National Center for Air Pollution Control were more responsive than they had been before the Senate committee. John Middleton explained that it had taken his office more than three years to get organized and "to understand what criteria should be." But the problem was not entirely an organizational one. The Public Health Service, through its Division of Air Pollution and later the National Center for Air

Pollution Control (NCAPC), had allocated few resources to the development of criteria. Initially only three professionals worked full time on criteria development, and by October 1966, only two people did the work. Of these two, one worked on sulfur oxides. At the time of the House hearings in August 1967, NCAPC had increased the number of professionals to seven, but they were now developing criteria for at least five air pollutants. After the hearings, NCAPC indicated that it would expand the staff assigned to criteria development to thirty-five by the end of the fiscal year on June 30, 1968.

There is strong reason to believe that the projected fivefold increase in personnel came solely as a result of congressional pressure. In its budget hearings the previous April, NCAPC had asked for only seven additional people, who would focus both on criteria development and on clinical research.[26]

During the hearings, many groups that had appeared before Muskie's committee also appeared before the House Commerce Committee. Few of these groups changed either their tone or their apprehension about the first criteria document or the possibility of federally mandated emission standards. For example, one coal executive labeled as erroneous HEW's claim that SO_2 threatens the public's health.[27] Such complaints found a sympathetic response in the Commerce Committee, where West Virginia's Harley Staggers was chairman.

At the end of the hearings, the Commerce Committee's reaction to the initial criteria was similar to that of the Senate. This reaction acknowledged the importance of criteria and similar confusion about the purpose of air quality criteria. Consequently the Commerce Committee accepted the Senate's wording on criteria and technology documents without modification, but it did add a requirement that both documents "have a substantial evidentiary backup." By the time the full House voted on the proposed legislation, relatively few differences existed between its version and that of the Senate. The conference committee met and easily negotiated a compromise acceptable to both chambers. The Air Quality Act of 1967 became law on November 21, when President Johnson approved what one senator called "the greatest single step taken by any Congress so far toward eliminating filthy air."[28]

The new law represented incremental change and a gradual increase in the federal government's involvement in abating unacceptable levels of air pollution. Under the legislation, the federal government would continue to identify harmful pollutants, and the states and local governments would retain primary responsibility for abatement and enforcement. Now, however, the federal government was requiring the states to develop air quality standards. Once a state established standards, an implementation plan would have to assure attainment of the standards "within a reasonable time." HEW was directed to develop air quality standards and implementation plans for states that failed to comply, but the likelihood of federal involvement was minimal. In its report on the legislation, for example, the House Commerce Committee emphasized "that this residual power in the Secretary of HEW will seldom, if ever, be used, since the States are expected to take the necessary steps to establish and enforce air quality standards."[29] The law was restrictive of HEW's involvement in this one area, but the Air Quality Act allowed HEW to initiate independent enforcement proceedings if a state's failure to enforce its air quality standards led to non-attainment of the standards or if a state's lack of action led to "an imminent and substantial endangerment to the health of persons."[30] In the first situation, the law repeated the earlier requirement that courts balance consideration of health effects with the practicability and economic and technological feasibility of compliance.

Its advocates hailed the new Air Quality Act, but the legislative process had taken a heavy toll on the PHS and the National Center for Air Pollution Control. Because NCAPC's activities would eventually lead to regulation, several industrial groups would be adversely affected. Accordingly they could not be counted on to provide NCAPC with external support. As Francis E. Rourke has generalized, political support is the basis of an agency's power and in the absence of external support, good relations with legislators are especially important to an agency. Supportive legislators can act as spokesmen for an agency, rebuff attempts to diminish or redefine an agency's responsibilities, and attempt to ensure an agency's well-being.[31]

The congressional hearings on the Air Quality Act had demonstrated the limited legislative support for NCAPC's prior activities. In its rejection of the first criteria, the Congress had questioned the environmental agency's scientific expertise and credibility and had found both lacking. As Davies has pointed out, NCAPC and its parent agency, the PHS, had chosen a risky strategy. By portraying themselves as competent scientists, the officials in the PHS had hoped to shield their efforts from outside pressure. When the criteria document was accurately and legitimately criticized, however, the strategy collapsed, and the PHS was left defenseless.[32] As a result, the agency's scientific data became fair game for all critics. In rejecting the initial criteria, in mandating their substantial revision, and in requiring extensive outside consultation, the Air Quality Act reinforced this vulnerability and provided a clear signal that the government's air pollution officials would have to be responsive not only to the Congress and individual congressmen but to affected interest groups as well. If this observation is accurate, then the Air Quality Act serves as an excellent illustration of Lowi's interest-group liberalism, in which "the role of government is one of ensuring access particularly to the most effectively organized, and of ratifying the agreements and adjustments worked out among the competing [interest-group] leaders and their claims."[33]

Further Efforts to Succeed

The Air Quality Act's requirements delayed HEW's efforts to produce further criteria. The question was how long the delay would be. Surprisingly, NCAPC's

Middleton believed the delay would be minimal, and in January of 1968, he predicted that revised criteria for sulfur oxides would be published in the early summer. In the meantime, the coal industry continued unabated its criticisms of the need for criteria. In early 1968, trade journals were filled with editorials and advertisements on the issue of SO_2 and health. *Coal Age,* for example, doubted whether the pollutant is a health hazard "in the concentrations encountered by the average person." Research had supposedly found that concentrations of SO_2 as high as 2 parts per million are not dangerous, even to the unhealthy. (The initial criteria had recommended that levels of SO_2 not exceed 0.08 parts per million over a period of twenty-four hours.) Without identifying them, *Coal Age* also claimed that some government agencies conducting research had been unable to link chronic exposure to air pollution and any ill effects on humans.[34]

Such criticisms kept the question of air quality criteria in the forefront of environmental activity in 1968. Both the House and the Senate conducted oversight hearings during the year. Unlike the reaction it had received the previous year, however, the NCAPC received little criticism of its performance. More important, the members of the Senate Subcommittee on Air and Water Pollution acknowledged that criteria should be published even though a comprehensive 'knowledge of a pollutant's effects was still unavailable. This was a clear repudiation of the coal lobby's argument that publication of criteria should be delayed pending the availability of incontrovertible data. The National Coal Policy Conference was willing to accept only those data for which complete consensus existed. According to the American Petroleum Institute and the Coal Policy Conference, such consensus did not exist, and therefore it was still premature to say that a particular level of any air pollutant is dangerous. As Senator Howard Baker (R.-Tenn.) observed, however, "Responsible public policy cannot wait upon a perfect knowledge of cause and effect."[35]

Despite the Congress's seeming encouragement of further activity on the part of pollution control, a new document on sulfur oxides was still unavailable by mid-1968. Middleton, now commissioner of the newly created National Air Pollution Control Administration (NAPCA), was able to offer only a progress report. At year's end, he said that the members of the National Air Quality Criteria Advisory Committee had already met five times, and its subcommittee on sulfur oxides had consulted nearly twenty specialists outside the federal government. In addition, every cabinet-level department, the Federal Power Commission, and the Tennessee Valley Authority had been given an opportunity to review and comment on preliminary drafts of criteria for particulates and sulfur oxides. Outside consultation was so extensive that the *Air/Water Pollution Report* predicted that Middleton's agency would face little opposition when it released the revised criteria.[36] Even Middleton was hopeful that the prepublication review of the new criteria would curtail or possibly eliminate the same kind of conflict and embarrassment that had surrounded publication of the first criteria.

Such optimism was unwarranted. According to Davies, most members of HEW's advisory committee complained that their opinions on the new criteria were ignored.[37] Moreover a few weeks before the release of a preliminary draft of the revised criteria in December 1968, the Manufacturing Chemists Association charged that its members had not been given an opportunity to see the forthcoming draft. The National Coal Association and the National Coal Policy Conference had received a draft, but that did not diminish their opposition. These complaints were at least partially successful because the criteria's section on health effects was rewritten before final publication on February 11, 1969.

The new criteria for sulfur oxides were little changed from the earlier document. The new version contained a more extensive review of existing literature, including that published in the intervening twenty-two months, but some of the problems evident in the first effort were still to be found. The new document acknowledged that much was still unknown about the effects of sulfur oxides. It suggested, however, that as further information became available, it would probably "show that there are identifiable health and welfare hazards associated with air pollution levels that were previously thought to be innocuous."[38] For this reason, Middleton believed that a reassessment and, possibly, a revision of the criteria ought to take place at least every five years. Like its predecessor, the new criteria document also indicated that many health effects are not caused solely by SO_2 but rather result from the additive or synergistic reactions between sulfur oxides and other pollutants. Nonetheless the report summarized few studies focusing on these reactions, and the minimum safe levels recommended in the criteria were provided solely for SO_2 and not for combinations of air pollutants.

The fact that the criteria recommended minimum safe levels was controversial. At the last meeting of HEW's advisory committee in December 1968, officials from NAPCA had told the members that the document would not summarize the research to the point of indicating minimum acceptable levels. Middleton's superior, Charles C. Johnson, administrator of the Consumer Protection and Environmental Health Service within HEW, felt that the government had an obligation to provide the states with some conclusions on minimum safe levels. His position seemed to be consistent with the intention of the Air Quality Act that the secretary of HEW issue criteria "as in his judgment may be requisite for the protection of the public health and welfare." Thus Johnson rejected the advisory committee's understanding, overturned Middleton's earlier decision, and required that the following paragraph conclude the criteria:

On the basis of the foregoing information and data, it is reasonable and prudent to conclude that sulfur oxides of 300 $\mu g/m^3$ or more in the atmosphere over a period of 24 hours may produce adverse health effects in particular segments of the population. In the promulgation of ambient air quality standards, it should be recognized that circumstances existing within a given region (as well as requirements for margins of safety) may warrant more stringent standards than those indicated by these criteria.[39]

Johnson's required statement on minimum safe levels rankled not only Middleton and his agency but the National Coal Policy Conference as well. The conference claimed that such a statement was equivalent to setting national air quality standards, and this was contrary to the intent of the Air Quality Act. Congressional pressure and adverse reaction from trade and industrial groups caused Johnson to change his mind. As a result, NAPCA issued a revised conclusion that deleted the suggestion that 300 micrograms per cubic meter of SO_2 represented the lower limits of acceptable air quality. The new conclusion noted that adverse health effects could be observed when levels of SO_2 exceed 300 micrograms per cubic meter for three to four consecutive days. Additionally the revised version softened its recommendation that 300 micrograms per cubic meter should be the upper limit for air quality standards. The report simply indicated that "it is reasonable and prudent to conclude that when promulgating ambient air quality standards, consideration should be given to margins of safety which would take into account long-term effects on health . . . occurring below the [stated] levels."[40]

The change did not deter further criticism. *Coal Age* and *Electrical World* printed editorials suggesting that the new criteria for sulfur oxides were scientifically invalid. As an illustration, *Coal Age* claimed that no evidence existed to indict sulfur oxides at the concentrations found in most American cities. There were even some people who felt that the new criteria were no better than the first effort. Typical of this position was that of Eric J. Cassell, a member of the criteria advisory committee, who contended that the new document was too one-sided to serve as anything more than a political document.[41]

In at least one respect, the harshness of the criticisms seemed inappropriate. Comparison of the data in table 4-1, which summarizes some of the new criteria's findings, with the data from the 1967 criteria reveals some interesting differences. According to the 1969 criteria, adverse health effects could be expected when annual levels of SO_2 exceeded 0.04 parts per million. In the first criteria, annual levels above 0.015 parts per million were said to be detrimental to certain groups. Levels of SO_2 as high as 0.04 parts per million over the same time period were said to be in the "range of concentrations and exposure times in which deaths have been reported in excess of normal expectation."[42] No explanation was provided for the difference, but it was certainly one that the coal and oil industries could appreciate.

Although the new criteria were not free of criticism, the complaints were insufficient to bring about any large-scale overhaul. More than five years after the Congress had first required publication of criteria, a document had finally been issued that did not have to be reworked. The document on control technologies had also been published so the states could proceed to set their own air quality standards.

Table 4-1
Resume of 1969 Air Quality Criteria for Sulfur Oxides

| | Level of SO_2 | | |
Time Period	$\mu g/m^3$	ppm	Effect
Twenty-four hours	>300	0.11	Adverse health effects if levels maintained for three to four days
One year	>115	0.04	Adverse health effects
One year	85	0.03	Adverse effects on vegetation
One year	345	0.12	Adverse effects on materials

Source: U.S. House of Representatives, Committee on Interstate and Foreign Commerce, Subcommittee on Public Health and Welfare, hearings, *Air Pollution Control and Solid Wastes Recycling, Part I,* 91st Cong., 1st and 2d sess., 1969–1970, p. 28.

Conclusions

As the 1960s ended, the proponents of clean air seemed to have the necessary tools to achieve their goals. The Air Quality Act put the federal government in the position of requiring state action on air pollution, and the government partially fulfilled this responsibility by publishing revised criteria for sulfur oxides. The criteria, while still controversial, could serve as the basis for implementation of the nation's air pollution program. The likelihood of this success appeared to be promising. The states would be setting their air quality standards at the same time that millions of Americans were demonstrating a growing concern for increased levels of environmental protection. Indeed environmental quality would soon become one of the most popular of all domestic policies, and few policy goals would have as much public support as cleaner air.

The Air Quality Act created a mechanism to channel this support into the standard-setting process. The law required all states to hold public hearings before they established air quality standards, and the large number of participants in many of the initial hearings was unprecedented. Most citizens shared a common goal—they wanted high levels of air quality. The major issue, of course, was not the extent of public participation but its impact. Could this support and enthusiasm for acceptable levels of air quality be translated into substantive achievements?

Notes

1. The President's Message, "Protecting Our National Heritage," is reprinted in U.S. House of Representatives, House Document 47, 90th Cong., 1st sess., 1967.

2. Ibid.

3. U.S. House of Representatives, Committee on Interstate and Foreign Commerce, hearings, *Air Quality Act of 1967,* 90th Cong., 1st sess., 1967, pp. 572–585.

4. Ibid.

5. "Controversy Developing Over SO Criteria," *Air and Water News* 1 (April 10, 1967):2.

6. Cited in House, *Air Quality Act of 1967,* p. 571.

7. U.S. Senate, Committee on Public Works, Subcommittee on Air and Water Pollution, hearings, *Air Pollution–1967, Part 3,* 90th Cong., 1st sess., 1967, p. 2041.

8. "Air Pollution Control," *Weekly Compilation of Presidential Documents* 3 (1967):651–652.

9. Senate, *Air Pollution–1967, Part 2,* p. 762.

10. Ibid., *Part 4,* p. 2514.

11. Ibid., *Part 2,* p. 780.

12. Ibid., *Part 4,* p. 2260.

13. Ibid., *Part 3,* p. 1117.

14. Ibid., p. 1140.

15. U.S. Senate, Committee on Public Works, Report No. 90-403, *Air Quality Act of 1967,* 90th Cong., 1st sess., 1967, p. 10.

16. Randolph's successes are reported in J. Clarence Davies III, *The Politics of Pollution* (New York: Pegasus, 1970), p. 57.

17. Senate, *Air Pollution–1967, Part 3,* p. 1142.

18. Dean Schooler, *Science, Scientists, and Public Policy* (New York: The Free Press, 1971), p. 44.

19. Senate, *Air Quality Act of 1967,* p. 3.

20. S780, 90th Cong., 1st sess., 1967.

21. Senate, *Air Pollution–1967, Part 3,* p. 1466.

22. Senate, *Air Quality Act of 1967,* p. 27.

23. Ibid., p. 28.

24. Ibid., p. 70.

25. House, *Air Quality Act of 1967,* pp. 163–164.

26. U.S. House of Representatives, Committee on Appropriations, hearings before a subcommittee, *Departments of Labor and Health, Education, and Welfare Appropriations for 1968,* 90th Cong., 1st sess., 1967, p. 413. By January 1, 1970, the unit with responsibility for criteria and standards consisted of only fourteen professionals, far fewer than the promised level of thirty-five.

27. House, *Air Quality Act of 1967,* p. 612.

28. *Congressional Record* 113 (November 14, 1967):32478.

29. U.S. House of Representatives, Committee on Interstate and Foreign Commerce, Report No. 90-728, *Air Quality Act of 1967,* 90th Cong., 1st sess., 1967, p. 12.

30. The Air Quality Act of 1967, PL 90-148, sec. 108, November 21, 1967.

31. Francis E. Rourke, *Bureaucracy, Politics, and Public Policy,* 2d ed. (Boston: Little, Brown, 1976), chap. 3.

32. J. Clarence Davies in comments to the author, August 1978.

33. Theodore J. Lowi, *The End of Liberalism* (New York: W.W. Norton, 1969), p. 71.

34. "The Sulfur Question," *Coal Age* 73 (April 1968):71. For similar criticisms, see *Electrical World* 169 (January 1, 1968):51.

35. U.S. Senate, Committee on Public Works, Subcommittee on Air and Water Pollution, hearings, *Air Pollution—1968, Part 2,* 90th Cong., 2d sess., 1968, p. 528.

36. *Air/Water Pollution Report* 6 (November 4, 1968):365.

37. Davies, *Politics of Pollution,* p. 161.

38. HEW, NAPCA, *Air Quality Criteria for Sulfur Oxides* (1969), p. iii.

39. Cited by *Air and Water News* 3 (February 17, 1969):3. Much of the discussion that follows relies on Davies, *Politics of Pollution,* pp. 161-162, and John C. Esposito, *Vanishing Air* (New York: Grossman, 1970), pp. 283-286.

40. HEW, *Criteria for Sulfur Oxides,* p. 162.

41. U.S. Senate, Committee on Public Works, Subcommittee on Air and Water Pollution, hearings, *Implementation of the Clean Air Act Amendments of 1970, Part 2 (Title I),* 92d Cong., 2d sess., 1972, p. 492.

42. HEW, PHS, *Air Quality Criteria for Sulfur Oxides* (1967), p. liii.

5 The Third Time Around

It did not take long for dissatisfaction with the Air Quality Act to arise. What had been viewed as an appealing compromise in late 1967 was widely acknowledged to be a bureaucratic morass less than two years later. For many people, the law was a failure. Perhaps the best indication of this belief came from a retrospective analysis of the law's impact. Speaking in early 1971, William D. Ruckelshaus, the first administrator of the EPA, commented that "not one grain of dust, not one liter of gaseous pollution has yet been removed from the atmosphere of this Nation as a direct result of the 1967 air legislation."[1] Because the Air Quality Act had been so thoroughly hailed at its passage, it is important to examine some of the reasons why the law failed to bring about the acceptable levels of air quality that it had seemed to promise.

One of the major reasons for failure was that the Congress had devoted little attention to the mechanics of implementing the law's requirements. The law, like its predecessor, the Clean Air Act, had placed the major responsibility for implementation with the states. In this regard, the provisions of the law were similar to those of many other laws passed in the 1960s. James L. Sundquist has pointed out that prior to 1960, many federal programs were intended to assist state and local governments in achieving their own goals.[2] The implication was that the national government would assist lower levels of government that formulated their own policies and implemented their own programs.

Sundquist claims that a major change occurred after 1960. Supposedly federal programs (and their associated grants of financial aid) became a means of accomplishing national objectives that were formulated in Washington, D.C., not in the individual states. According to this new arrangement, state and local governments would implement these new programs as a matter of administrative convenience for the federal government. The nationalization of policy goals has its advocates, but the strategy is not without its difficulties. When congressmen pass laws, they can ignore the structural problems inherent in federalism, but bureaucrats cannot. When they attempt to implement congressional policies that require the states' cooperation, bureaucrats must hope that the states, which are legally independent, will cooperate. Federal administrators, however, have no effective way to discipline defiant state officials who disagree with the national objectives or who believe that implementation should be handled differently. Such an arrangement calls for a great deal of patience and coordination among all parties, both state and federal. For the state-federal arrangement to succeed, the complete cooperation of the states is necessary.

When the Congress passed the Air Quality Act, it had not considered that intergovernmental coordination might be a problem. As a result, the law had not given HEW sufficient authority to compel either compliance or coordination. Thus the accomplishment of the nation's goal of clean air rested less on HEW's own efforts than it did on the good offices of fifty state governments. One crucial consequence of giving the states primary responsibility for implementation was that it brought about the separation of policy formulation from policy implementation. Under the terms of the Air Quality Act, a single set of goals and guidelines would be developed for all states, regardless of differences among them. At the state level, administrators would implement policies with which they might have little or no allegiance. These officials would be expected to implement other people's ideas without any discussion of political or organizational feasibility. Many studies of public policy emphasize that a sure way to reduce a program's effectiveness is to separate its formulators from its implementors.[3]

A further indication that policy formulation should not be separated from implementation comes from the realization that the Air Quality Act created an inordinate number of points at which implementation could be impeded. In Pressman and Wildavsky's terms, the number of separate clearance points was excessive. A clearance point is simply an instance in which an individual participant is required to give his consent.[4] As the number of clearance points increases, the likelihood of successful implementation decreases, or so Pressman and Wildavsky have argued. In their analysis of the operation of a federal program in Oakland, Pressman and his coauthor found that relatively few clearance points had effectively ruined "laudable intentions, commitment, and an innovative spirit."[5]

The same problems were evident in the Air Quality Act. The fact is that the number of clearance points in the Oakland project pales in comparison with the number of such points in the Air Quality Act. For a national air quality program to be established for only one pollutant, more than a thousand clearance points would have to be passed. As an example, the law required the secretary of HEW to consult with state and local authorities before the designation of each air quality control region. Next the governors would notify HEW of their intent to set air quality standards for each of the projected 247 regions. Each point provided an opportunity for delay and ensured the need for negotiation among polluters and state and local officials.

The significance of the clearance points can be seen by examining the law's actual implementation. The Air Quality Act required that all control regions be established no later than May 1969, but by February of that year, HEW had issued criteria and control-technology documents for only two pollutants (particulates and sulfur oxides) and had designated only 6 regions (the metropolitan areas of Denver, Chicago, Philadelphia, Los Angeles, Washington, D.C., and New York City). The slow pace in designating regions guaranteed lengthy delays. No action was expected from the states until after regions had been designated, but at the rate of 6 regions in approximately eighteen months, it would take HEW more than sixty years to designate all of the regions. It is not surprising, therefore, that more than two years after the passage of the air pollution law, the program had few achievements to its credit and no approved state implementation plans.

The Air Quality Act did have structural faults, but the problem was not entirely with the law. Also at fault were ineffective bureaucratic efforts and the states' incapacity to respond. In terms of organizational efforts, for example, rhetoric on the importance of pollution control far exceeded action during the 1960s. Rourke has suggested that the ability of an agency or department to resist reorganization reflects the strength of its constituent support and, therefore, its likely effectiveness.[6] If this argument is correct, then it is worthwhile to review briefly the status of air pollution control efforts at the national level during the 1960s.

HEW and its subordinate agency, the PHS, had overall responsibility for air pollution in the 1960s. Within the PHS, however, the Bureau of State Services contained a Division of Air Pollution after the fall of 1960. The fact that the division was subordinate to so many other units reveals the relative unimportance of antipollution efforts at the time. In 1966, the division became the National Center for Air Pollution Control. The change gave the unit some autonomy and greater visibility, but, as Davies has written, it "still suffered from an acute case of bureaucratic layering."[7] It was far removed from the secretary of HEW and had to work through several offices before reaching the department's highest officials.

The National Center for Air Pollution Control underwent an internal reorganization in mid-1967 that consolidated all activities relating to the development of air quality criteria into a single unit. A year later, still another reorganization occurred; this one made the rechristened National Air Pollution Control Administration (NAPCA) one of three units in a newly created Consumer Protection and Environmental Health Service, which was subordinate to an assistant secretary in HEW. According to this secretary, the creation of NAPCA meant that the government's air pollution efforts had been upgraded to a status "fully commensurate with the magnitude and importance of the contemporary air pollution problem."[8] This argument was unconvincing to at least some members of Congress. In at least one report, members of a House committee had warned of their dissatisfaction with the impending changes as well as the status of some of the government's environmental activities.[9] Later analyses by other observers confirm the existence of organizational problems after NAPCA's creation. In his book, *Clean Air,* for example, Charles Jones notes that NAPCA floundered so much that Secretary of HEW John Gardner had to create a special high-level committee to guide the agency's programs.[10] The election of a Republican president in 1968 did not improve NAPCA's effectiveness or even alter relations among its top officials. Relations among some of these officials became so poor that the agency's assistant commissioner reportedly told the secretary of HEW in 1970 that NAPCA's head, John Middleton, was engaged in "a deliberate, long-range, carefully calculated plan to eliminate any effort to execute an effective air pollution program."[11]

The organizational problems inherent in the nation's air pollution efforts were a prime cause for the establishment of the EPA in late 1970. President Nixon had expressed his concern for existing arrangements shortly after he took office when he asked the Advisory Council on Executive Organization (commonly known as the Ash council for its chairman, Roy Ash) to consider possible reorganization and consolidation of all federal environmental programs. This request, made in December 1969, was followed two months later by the president's public acknowledgment of organizational deficiencies among the government's pollution control agencies.[12] By April 1970, the Ash council had recommended the creation of a comprehensive environmental agency, which would include programs from fifteen different existing agencies responsible for pesticides, solid waste, environmental radiation, and air and water pollution.

Congressional assent was necessary for the proposed reorganization. During congressional consideration of the plan, what was noteworthy was not the lack of opposition but the extent of enthusiasm for the change. Most congressmen appreciated the problem of organizational and bureaucratic inefficiency and hoped that a new agency would be the solution.

The chairman of the CEQ, Russell Train, was the administration's chief spokesman for reorganization. He assured the congressmen that the new agency would solve problems of coordination and that everyone involved in the change was enthusiastic about the prospects of working in a consolidated agency.[13] As one analysis has suggested, however, there was little reason to be so optimistic. The name of the agency would be new, but the EPA would be constructed from "the same organizational units that had been publicly judged to be inadequate in the late 1960s."[14] Thus much of the enthusiasm attached to the EPA's creation in December 1970 appeared to be inappropriate. To a large extent, the EPA has not been able to overcome some of its early problems. Many studies continue to document the agency's operating and organizational difficulties, problems that have had a substantial impact on the agency's programs and its ability to define safe levels of air quality.[15]

The final problem with implementation of the Air Quality Act was the general inability of the states to comply with the law's requirements. Appropriate emission standards can be set only when adequate information is available on the location of major polluters and on the volume, frequency, and duration of their emissions. Such data can be gathered only if sufficient staff and adequate monitoring devices are available. For the most part, few states had either during the late 1960s. One indicator of the states' abilities was the poor quality of their implementation plans submitted to HEW and NAPCA. Rarely were plans submitted on schedule, and of the first twenty sent to NAPCA, many deficiencies were identified, ranging from inadequate control regulations to a lack or poor description of resources. Partial approval of the first implementation plan was granted in the summer of 1970, nearly three years after passage of the Air Quality Act. In short, these many deficiencies revealed that substantial remedial action would be necessary if the nation's efforts to achieve clean air were going to become anything more than a symbolic commitment to environmental quality.

Drastic Reform: The Clean Air Amendments of 1970

The House Subcommittee on Public Health and Welfare wasted little time in its consideration of the inadequacies of the Air Quality Act. Congressman Paul Rogers (D.-Fla.), the second-ranking member of the subcommittee, initiated two days of oversight hearings in December 1969, a full seven months before the law's authorizations would expire. Much of the early testimony concerned the Nixon administration's evaluation of progress under the act. Officials from the PHS and NAPCA were pleased with their efforts and accomplishments, but this was not the case for many representatives, especially Rogers. He continually charged that existing efforts were inadequate and that rapid progress was necessary.

After a break of several months, the committee reconvened its hearings in March 1970. By this time, President Nixon had delivered a message to Congress on environmental quality that made clear his discontent with existing legislation and his desire for widespread improvements. Nixon criticized the lengthy delays built into the Air Quality Act and then proposed that the federal government "establish nationwide air quality standards, with the states to prepare within one year abatement plans for meeting those standards."[16] The purpose would be to provide a minimum standard for acceptable levels of air quality and to relieve the states of the delays related to their own standard-setting processes. The proposal meant that the federal government would now define acceptable levels of air quality, but Nixon's message did not indicate how this process would occur.

The specific plans on how to establish national air quality standards were contained in a House bill that was introduced to coincide with Nixon's speech. The administration bill would require the secretary of HEW to publish, within six months and after consultation with advisory committees and other federal agencies, "proposed regulations establishing nationally applicable standards of ambient air quality for any pollutant or combination of pollutants which he determines endanger or may endanger the public health or welfare."[17] After consideration of public comments, the secretary would "promulgate such modifications as he deems appropriate." Once the national air quality standards were established, the states would retain their right to set more stringent standards. The states would also maintain primary responsibility for setting emission standards for criteria pollutants, for developing implementation plans, and for enforcing the law.

The House of Representatives approved a bill that paralleled the administration's.[18] Both required states to achieve acceptable levels of air quality "within a reasonable time"; both agreed that publication of a document on recommended control techniques could follow (rather than precede) the setting of air quality standards; and both required more rapid designation of control regions. There were also some important differences. First, unlike the Nixon plan, which had not mentioned criteria, the House required that criteria be the basis of national air quality standards. Second, the House did not require state implementation plans to utilize emission standards, whereas the Nixon proposal did. Third, the House required the secretary of HEW to publish proposed air quality standards (for pollutants with criteria already issued) within thirty days of the law's passage; Nixon's bill had allowed six months. Fourth, though the administration preferred to delete it, the House retained the admonition to the courts that they should give "due consideration to the practicability and to the technological and economic feasibility of complying" with the law in enforcement actions brought before them.

At the same time that the House Subcommittee on Public Health and Welfare was considering revisions in the 1967 law, the Senate Subcommittee on Air and Water Pollution was also conducting hearings on the same topic. In addition to Nixon's proposal, the committee was also examining Senator Muskie's proposal, a bill designed to improve and speed up the implementation of the Air Quality Act of which Muskie had been a major author. In his proposed revisions, Muskie favored continued reliance on state initiative, avoided national air quality standards, and required implementation plans to include emission standards for all sources of air pollution. With the exception of these few changes, Muskie's bill was much like the existing legislation. His solutions to the many problems of the Air Quality Act were primarily remedial in nature.

This approach was soon to be viewed as unsatisfactory. President Nixon's ideas were far more innovative, and in proposing them, he had successfully outmaneuvered Muskie, a potential presidential opponent. The president's initiative put the senator in a defensive position; Muskie would have to alter his stance if he had any hope of retaining leadership in the environmental arena.

Much of the Senate subcommittee's discussion focused on the possible role of criteria under the Nixon plans, but several senators were still confused about their purpose and the relation between them and air quality standards. After seven years of discussing the two concepts, even Muskie had to ask if standards were different from criteria (though his own bill relied on the distinction between the two).[19]

The confusion increased during a special executive session of the subcommittee in which Commissioner Middleton of NAPCA tried to explain the rationale for national air quality standards. Middleton said that such standards would ensure that each area's air quality would be at a level to be protective of the public's health; that is, the standards would protect against a minimum adverse health effect. When asked if national standards would be set at a level at which no adverse health effects occurred, Middleton responded that they could, but that such a level would be nearly impossible to specify.[20] To remedy this problem, a margin of safety would be used to discriminate between the level at which health effects are first noticed and a level considered to be acceptable. Muskie was puzzled; if national standards were intended to avoid all adverse effects, why would the administration bill allow states to set more restrictive standards? Middleton answered that cases could exist "in which standards for protection of health may not guard against environmental and economic effects."[21] Thus national standards might not be protective of the public's welfare. Each state would have to decide if it desired its own, more stringent standards that would be protective of plants, animals, and materials. Middleton also said that national standards would be likely to reflect possible synergistic reactions among pollutants. As a consequence, Middleton gave the impression that if two or more pollutants interacted to produce harmful effects at levels lower than any of the pollutants individually, then the national standard would be set to protect against these synergistic effects.

A further question involved the time period in which the states would have to achieve the national standards. Here again, Middleton was imprecise. He emphasized that the standards, as goals, would have to be met within "a reasonable time" (both the Nixon and Muskie plans had retained the Air Quality Act's reasonable-time requirement). This period, observed Middleton, "is not an awfully elastic time. It has got to be as soon as it can be within the present technology and resources available."[22] Upon further questioning, he conceded that the states would have to establish reasonable deadlines in their implementation plans. Middleton then listed more than a dozen variables that could influence the time limits, including a state's priorities and concentrations of industry, the availability of clean fuels and control technology, and the potential economic impact of a state's control program.[23] In short, Middleton seemed unwilling to commit himself to any specific deadlines.

A consistent underlying theme of Middleton's testimony reflected the belief that the states should have prime responsibility for making decisions affecting their citizens. With hundreds of new federal programs established in the 1960s under Presidents Kennedy and Johnson, the states had been bypassed frequently and, in some instances, ignored. President Nixon's concept of a New Federalism attempted to reverse this trend. First outlined during the summer of 1969, the plan envisaged what Nixon called "responsible decentralization," which meant a move away from federal control back to state and local initiatives. Nixon was not opposed to the national government's providing financial assistance to lower levels of government, but he did stand against a multitude of categorical grants that severely limited the flexibility of these lower levels. A major part of the New Federalism called for the states to have broad discretion in the implementation of national goals. Thus Middleton's plea for such discretion was understandable in light of Nixon's philosophy.

Regardless of Middleton's consistency with the New Federalism, his responses were unacceptable to Senator Thomas F. Eagleton (D.-Mo.), who complained that the mere establishment of a national standard without a mandated deadline amounted to a "pontifical goal, that is baloney, too."[24] Eagleton's concern was well founded. According to the provisions of the Nixon, Muskie, and House proposals, attainment of acceptable levels would remain an elusive goal. NAPCA was still planning the release of many more criteria. After their publication, national standards, state standards, and implementation plans would follow for each new pollutant. Moreover, Middleton had indicated that the states should consider many intervening variables before setting any deadlines. Not only would an adequate consideration of these variables be time-consuming, but each variable could be used to argue for extensions of possible deadlines or from exemptions from deadlines once they were set.

By the time the Subcommittee on Air and Water Pollution concluded its hearings at the end of May 1970, it was clear that much dissatisfaction existed with all proposals under consideration, including Muskie's. Several things happened both to intensify this dissatisfaction and to guarantee major change. First, the Senate Committee on Public Works had not even issued a report on the clean air legislation, but the House of Representatives was less than two weeks away from passage of its own bill. In this instance, the House had seized the initiative at the expense of the Senate's chief environmental supporter, Edmund Muskie. Second, the country's first Earth Day in April had demonstrated an unprecedented degree of public support for environmental quality. Once the House acted, the burden would be on the Senate to respond to the enthusiasm that Earth Day had generated. It was unlikely that Muskie's original proposal would be satisfactory because it was simply an attempt to improve the Air Quality Act. The Public Works Committee could approve the Nixon bill with minor changes, but this would be politically embarrassing to the Senate's Democratic majority as well as personally embarrassing to Muskie. After his bid for the vice-presidency in 1968, the senator was widely considered to be the top prospect for his party's presidential nomination in 1972. Thus to approve the Nixon proposal would be tantamount to conceding Muskie's major strength—the environment—to his probable opponent.

Third, Ralph Nader's Task Force on Air Pollution had clobbered Muskie in early May with the release of *Vanishing Air*.[25] This scathing critique of the nation's clean air programs blamed Muskie for nearly every problem that existed. The senator was accused of providing inadequate support for NAPCA, of ignoring the deficiencies of the Air Quality Act, and of failing to hold oversight hearings on NAPCA's activities. The Nader report even suggested that Muskie resign his subcommittee chairmanship because of his alleged incapacity to operate in favor of the environment.

These events pressured Muskie to recoup his reputation. A dramatic move was in order, and such a move came in mid-August when the Subcommittee on Air and Water Pollution reported its proposal for a new clean air law to the full Committee on Public Works. In terms of acceptable levels of air quality, the committee proposed the following:

1. More rapid publication of intended criteria.
2. Utilization of all criteria previously published. These documents would be subject to periodic review and modification.
3. Publication of national air quality standards within four months for all pollutants for which air quality criteria had already been issued. The standards would be set solely on the basis of health considerations. (In its discussion of the Air Quality Act in 1967, the Senate's report had emphasized that the development of air quality standards should be based on health, social, economic, and technical considerations.)[26]
4. Elimination of the reasonable-time requirement for compliance with the national air quality standards. In its place, states would have to achieve acceptable levels of air quality within three years from the date of approval of its implementation plan. Under the proposal's timetable, states would achieve acceptable levels of air quality for certain pollutants by mid-1975, but a one-year extension could be granted under some precisely defined conditions. The inclusion of a specific deadline meant that no longer would there be any justification for allowing economic or technological feasibility to preclude achievement of clean air. This requirement marked a clear repudiation of New Federalism and Middleton's plans. Instead of allowing the states flexibility and discretion, Muskie's committee specified a precise compliance date and curtailed the variables that states could consider in designing their implementation plans.
5. Elimination of the requirement that courts consider practicability and technological and economic feasibility in enforcement actions.
6. Elimination of the stipulation that publication of a document on recommended control techniques precede the development of implementation plans. Such documents would provide information only, not recommendations; further, the documents would no longer be required to discuss economic feasibility or cost-effectiveness analysis.
7. Affected parties could seek judicial review of air quality and emission standards within thirty days of their promulgation or "after such date whenever it is alleged that significant new information has become available."[27]

Nearly a dozen companies and industrial groups complained to the Committee on Public Works that parts of the proposal were either unfair or impossible to attain. As an illustration, the reasonable-time requirement had still been in the subcommittee's bill when its members had finished their hearings, but this requirement had been eliminated entirely during the summer. According to one report, Muskie added the deadline section shortly before the final proposal was released in August.[28]

Despite the complaints, the Committee on Public Works made few changes in its proposal. The committee's report on the bill left little to question; health would now be the sole determinant of air quality standards. The Senate report declared that polluters would either meet the law's requirement "or be closed down."[29] Nonhealth variables would be of minimal importance in achieving acceptable levels of air quality. For existing stationary sources of pollution, arguments about feasibility were rejected soundly. Although earlier legislation had intended that polluters do only what was technologically and economically possible, the new bill largely ignored such questions.

In addition to emphasizing health considerations, the Public Works Committee's report went a step further and provided specific guidelines on whose health national air quality standards ought to protect. In brief, the committee stressed that such standards should protect the health of any group in the population, including those people who are particularly sensitive to air pollution. The report emphasized that the standards should be set even lower than was necessary to be protective of the public's health. By introducing a margin of safety, "a reasonable degree of protection [would be] provided against hazards which research has not yet identified."[30] To fill gaps in existing knowledge, the Senate bill mandated more research on the relation between air pollution and its effects on health and welfare. Although existing criteria would be the basis of the first set of air quality standards, the report noted that the criteria would require occasional revision to ensure currency. Unlike its reaction in 1967, the Senate was only suggesting that some improvements in the criteria might be in order. It seemed that the delays inherent in the Air Quality Act were sufficient to convince the senators that further delay was untenable.

On September 22, 1970, after two days of floor debate and only minor amendments, the Senate voted unanimously for the product of Senator Muskie's subcommittee. Nine Senate and five House conferees began their deliberations on October 8 and ended them on December 16. The conference committee accepted the Senate's proposal with only a few modifications. In fact, the conferees from the House acquiesced to the Senate version in nearly every instance. The Senate conferees had successfully upheld the elimination of the reasonable-time requirement, the issue of feasibility for all existing stationary sources, and the courts' consideration of "practicability" in enforcement actions. At least two important changes are worth noting. First, the conference committee recognized the recent creation of the EPA by providing its administrator with responsibility for implementing the law. This official would be authorized to grant a two-year extension to the 1975 deadlines for meeting air quality standards, but the conditions under which he could do so limited his discretion.

Second, the conferees made a distinction between national air quality standards for public health and those for public welfare. Primary air quality standards would be "standards the attainment and maintenance of which in the judgment of the Administrator, based on such criteria and allowing an adequate margin of safety, are requisite to protect the public health." The deadline for achieving the primary standards in all control regions would be mid-1975. Secondary air quality standards would "specify a level of air quality the attainment and maintenance of which in the judgment of the Administrator, based on such criteria, is requisite to protect the public welfare from any known or anticipated adverse effects associated with the presence of such a pollutant in the ambient air."[31] Public welfare would include effects on crops, animals, climate, vegetation, and man-made materials, among others. Unlike their primary counterpart's mid-1975 deadline, secondary standards would have to be achieved within a reasonable time, which was left undefined.

Both the House and the Senate accepted the conferees' recommendations without change, and President Nixon signed the Clean Air Act Amendments on December 31, 1970. The new law, proclaimed Muskie, "carries the promise that ambient air in all parts of the country shall have no adverse effects upon any American's health."[32]

His promise was one that no one could guarantee. In mandating the use of national air quality standards and the immediate designation of all control regions, the Congress eliminated many of the clearance points that had delayed progress under the Air Quality Act. At the same time, however, the new law represented a distinct break with the past. No longer would polluters be allowed to tie reductions in their emissions to volunteerism or to the availability of control technology. In what Jones has called "spculative augmentation," passage of the Clean Air Amendments of 1970 marked congressional response to public demands for cleaner air at the expense of considerations of feasibility.[33] In other words, cleaner air was promised with only scant consideration given to the administrative and technological capabilities required to achieve it. In short, this situation created some very interesting possibilities, and one of these involved the EPA's first major requirement under the new legislation: the publication of national ambient air quality standards for SO_2.

Setting National Air Quality Standards

According to the Clean Air Amendments, the EPA's administrator would have thirty days after enactment to publish proposed primary and secondary air

adequately protected if twenty-four-hour levels do not exceed the standard of 365 micrograms per cubic meter, but if higher concentrations over shorter periods have an adverse effect, then a twenty-four-hour time period is too long. It may be the case that high exposures over a very short period will have an adverse effect on health or welfare, while exposure to the same amount over twenty-four hours will have no noticeable impacts. The EPA was at least partially responsive to people concerned with very short-term effects; in late March 1971, the agency added a secondary standard of 1,300 micrograms per cubic meter, which could not be exceeded over a three-hour period more than once a year.

For chronic effects, the use of an annual average suggests that no damage occurs if people, animals, or materials are exposed to low levels of sulfur oxides for periods longer than a year. Such an assumption is invalid, however, if sulfur oxides have cumulative, long-term effects that are not adequately prevented by an annual standard. Epidemiological studies could potentially identify these effects, but research of this sort is expensive and difficult to conduct over extended periods.

As far as unanticipated effects are concerned, researchers in all disciplines usually seek results they expect to find. In air pollution research, the general tendency is to look for such noticeable effects as coughing, shortness of breath, and so forth. When effects are not readily discernible or have long latency periods, they are less likely to be the subject of research. The criteria (and therefore the standards for SO_2) revealed just such a deficiency. The criteria included no discussion of a possible linkage between exposure to sulfur oxides and cancer, birth defects, or genetic mutations.

Third, some concern existed about the manner in which the standards included consideration of synergistic effects. To the extent that the standards were based on epidemiological research, they incorporated synergistic reactions. But the recognition that sulfur oxides can be involved in synergistic reactions does not mean that one can attribute ill effects to one pollutant or the other, and this has important implications. If the harmful effects attributed to sulfur oxides are actually due to sulfur oxides in combination with other pollutants, then sulfur oxides by themselves can still be harmful, but at levels far above the air quality standard or can be harmless (sulfur oxides can serve as catalysts leading to adverse reactions). In either case, efforts to control and abate SO_2 are potentially misdirected and possibly unfair to polluters.

Finally, there were complaints that the standards were too stringent. Such claims came primarily from industrial groups that would feel the standards' eventual impact in the form of emission limitations. The Kennecott Copper Corporation, whose smelters are a major source of sulfur oxides, was especially perturbed. Kennecott attacked the air quality criteria as representiing "extrapolations to general conclusions" and as reflecting "improper conclusions based on irrelevant data."[37] In short, it believed that the criteria were largely irrelevant to the standard-setting process. Such attacks as Kennecott's are entirely consistent with expectations about reactions to regulatory policies. When regulatory constraints are about to be imposed, groups subject to regulation are likely to question the scientific basis for the constraints.

Complaints about the standards also came from a rather unexpected source—one of the three members of the President's Council of Economic Advisors. This group has an advisory role, but the views of its members frequently parallel the president's. Hendrik S. Houthakker's speech to a business group a few days before final promulgation of the standards must have come as a surprise to the EPA. The Clean Air Amendments had prohibited the EPA from considering social or economic variables in setting the standards, and surely Houthakker knew this. Nonetheless he maintained that such an omission was unconscionable. Air quality standards, he offered, imply "a value judgment on the social importance of clean air relative to the social cost of achieving it." Accordingly he believed that some parts of the country should be allowed to relax the national standards if they wished to do so.[38] His speech indicated a misunderstanding of the need for and purposes of national standards, and more important, it presaged discontent with environmental programs within the Nixon administration. In the years to come, this discontent would seriously hinder the EPA's efforts to reduce levels of SO_2.

For the time being, however, Administrator Ruckelshaus had to issue the final national ambient air quality standards no later than April 30, which he did. With the exception of a short-term, secondary standard, which had been added in late March, the final standards were identical to those issued in January. Apparently the comments of both polluters and environmentalists had not altered the EPA's belief about the suitability of its choices. This meant that the final standards were not the product of extensive negotiation, as regulatory standards usually are. In avoiding the bargaining that usually occurs with affected interests, the EPA violated a prime tenet of administrative practice: the likelihood of cooperation is enhanced when affected parties are given an opportunity to participate in decision making.

When Ruckelshaus announced the final standards, he conceded that available scientific evidence on air pollution was imperfect, but he also emphasized that the need for further research "cannot justify failure to take action based on knowledge presently available."[39] Because considerable skepticism had been expressed about the feasibility of meeting the standards by mid-1975, Ruckelshaus also addressed that issue, though not to the satisfaction of the people concerned with it. He simply said that the EPA had ignored questions of feasibility. At the same time, however, Ruckelshaus did not believe that the issue of feasibility would cause problems in meeting the deadlines in most parts of the country. He predicted that polluters could comply with the law if they switched from high-sulfur coal to such low-sulfur fuels as residual fuel oil or natural gas. Ruckelshaus indicated that the use of natural gas across the nation would have to be increased by about 15 percent in order to attain clean air. The change to low-sulfur oil and natural gas would cause no major problems, inasmuch as Ruckelshaus implicitly acknowledged the widespread availability, albeit higher costs, of both fuels.

The belief that consumption of these fuels could be increased ignored several salient considerations. For natural gas, federal policy limited the likelihood of increased supplies. As a result of a Supreme Court decision in 1954, the Federal Power Commission was required to regulate the wellhead price of natural gas shipped across state lines. This regulation, which led to major differences in price between intrastate and interstate shipments, created an economic incentive to sell natural gas within the state in which it was produced since only the in-state customers could pay more than the interstate price established by the commission. Under these circumstances, few incentives existed to increase interstate shipments. In effect, federal policies encouraged the consumption of natural gas close to production sites (primarily in Texas and Louisiana), regardless of its potential environmental importance to other states.

For residual oil, Ruckelshaus ignored American refiners' apparent lack of interest in producing larger quantities of the fuel. By 1970, for example, the average yield of residual oil per barrel of crude oil produced in the United States had declined to 6.4 percent (from 8.1 percent in 1965). Equally important, Ruckelshaus seemingly ignored the major sources of the residual oil that were available. In the absence of adequate domestic sources, many electric utilities on the East Coast had become almost totally dependent on imported residual oil. Indeed almost 90 percent of the residual oil used in the eastern United States in 1970 was imported.[40] Much of the crude oil used to produce this residual oil came from the Middle East, the least desirable area from the perspective of national security.

The stringency of EPA's recommended limits on the allowable sulfur content of residual oil also had important ramifications. When the agency published proposed guidelines for the preparation of state implementation plans in April 1971, it indicated that regulations appropriate for meeting the new air quality standards would limit the sulfur in residual oil to 0.3 percent.[41] Prior to this recommendation, few states had required such low levels of sulfur content. In fact, such a level would be far lower than those to which many electric utilities were accustomed. Moreover if states adopted the 0.3 percent standard, oil-fired power plants would have to increase their consumption of fuel (because the heat value of residual oil, measured in Btu, decreases in rough proportion to the decrease in its sulfur content). Consequently when sulfur-in-fuel standards become more stringent, power plants have to burn more oil in order to maintain existing rates of production.

In addition to creating anxiety among users of residual oil, the EPA's plans for changes in the use of fuels also led to apprehension on the part of coal producers and users. The guidelines for implementation plans had recommended that certain states prohibit the combustion of coals with a sulfur content exceeding 0.7 percent. The EPA believed that the use of these coals would allow certain highly polluted areas to achieve safe levels of SO_2.

This assumption was probably correct, but it seems that little thought had been given to the feasibility of implementing such a strategy. On the one hand, the recommended level for coal was far lower than the coal used in previous years.[42] On the other hand, most of the country's coal-fired power plants are located in the East, and most of the country's recoverable low-sulfur coal reserves are located west of the Mississippi River. What the EPA seemed to ignore was the fact that western coals are expensive to transport to the East, are frequently incompatible with existing power-plant boilers, and are less efficient than eastern coals in their energy produced per pound.

However inappropriate the EPA's recommendations on coal, oil, and natural gas appear today, it must be remembered that the judgment of infeasibility is based on hindsight and on many events that could not be predicted in early 1971. At that time, the ability to implement a long-term plan was probably not a major concern to a new, energetic agency and administrator.

The EPA had gone on record as stating that success was possible and that all Americans would breathe clean air by mid-1975. Progress toward this goal would not be without its difficulties, but the likelihood of success would be enhanced if the new agency could count on the full support of President Nixon. His backing would demonstrate to other federal agencies that he expected their complete cooperation with the EPA's efforts. It would also indicate to affected private interests that they would not receive a sympathetic hearing when they complained about the agency's impact on their operations.

One would not have expected Ruckelshaus to accept his position with the EPA unless he had assurances of presidential support. Thus he must have been pleased at his swearing-in ceremony when the president told him that he would "have the support, completely, of the White House, of the President, and of this administration in his efforts to provide the programs which will bring clean air, clean water, and a better environment for all Americans."[43] After a few months in office, Ruckelshaus remained confident of the president's pro-environment position. Threatening to resign if the Nixon administration backed away from its commitment, Ruckelshaus declared war on polluters and claimed that the Nixon administration would not sacrifice a clean environment in favor of more jobs.[44]

Conclusions

Ruckelshaus and the EPA seemed to be in an enviable administrative position in early 1971. Charged with achieving the highly popular goal of clean air, they would be implementing a law that had received 374 of 375 votes in the House of Representatives and had passed without a single dissenting vote in the Senate.

And they would soon receive a 33 percent increase in funding for air pollution programs from the Congress. Finally the new agency was created at the request of a president who had devoted much of his first state of the union message to environmental protection and who had pledged his full cooperation and support to environmental programs.

All these assets appeared to guarantee success. There was little else for which the EPA could ask. In the words of one participant, the EPA seemed poised for great success not only in the area of air pollution but in all of its programs. According to John Quarles, who was one of Ruckelshaus's top assistants, the EPA's initial results were impressive. The continuation of such results were closely linked, however, to high levels of public support. Ruckelshaus had "tied the fortunes of EPA to public opinion as the only base of political support," and, as Quarles observed, "that dependence created a vulnerability; if ever the intensity of public support should wane, so would the effectiveness of the agency."[45]

Notes

1. Ruckelshaus's speech before the National Press Club in Washington, D.C., is cited by *Congressional Record* 119 (June 29, 1973):22296.

2. James L. Sundquist, *Making Federalism Work* (Washington, D.C.: The Brookings Institution, 1969), pp. 3, 12-13.

3. Edwin C. Hargrove, *The Missing Link* (Washington, D.C.: The Urban Institute, 1975), p. 29, and Jeffrey L. Pressman and Aaron Wildavsky, *Implementation* (Berkeley: University of California Press, 1973), p. 143.

4. Pressman and Wildavsky, *Implementation,* p. xvi.

5. Ibid., p. 6.

6. Francis E. Rourke, *Bureaucracy, Politics, and Public Policy,* 2d ed. (Boston: Little, Brown, 1976), p. 87.

7. J. Clarence Davies III, *The Politics of Pollution* (New York: Pegasus, 1970), p. 104.

8. The comments of Phillip R. Lee can be found in U.S. House of Representatives, Committee on Government Operations, hearings before a subcommittee, *Federal Air Pollution R.&D. on Sulfur Oxides Pollution Abatement,* 90th Cong., 2d sess., 1968, pp. 54-55.

9. U.S. House of Representatives, Committee on Science and Astronautics, Subcommittee on Science, Research, and Development, Committee Print, *Managing the Environment,* 90th Cong., 2d sess., 1968, pp. 35-36.

10. Charles O. Jones, *Clean Air* (Pittsburgh: University of Pittsburgh Press, 1975), pp. 111-112.

11. *Air and Water News* 4 (September 21, 1970):1-2.

12. See the President's Environmental Message to the Congress, reprinted in U.S. House of Representatives, House Document 225, 91st Cong., 2d sess., 1970.

13. U.S. House of Representatives, Committee on Government Operations, hearings before a subcommittee, *Reorganization Plan No. 3 of 1970*, 91st Cong., 2d sess., 1970, p. 20.

14. National Research Council, *Decision Making in the Environmental Protection Agency* (Washington, D.C.: NAS, 1977), p. 2.

15. See, for example, U.S. House of Representatives, Committee on Science and Technology, Subcommittee on the Environment and the Atmosphere, Committee Print, *Organization and Management of EPA's Office of Research and Development*, 94th Cong., 2d sess., 1976, p. 35.

16. House Document 225, p. 8.

17. HR 15848, 91st Cong., 2d sess., 1970.

18. HR 17255, 91st Cong., 2d sess., 1970.

19. U.S. Senate, Committee on Public Works, Subcommittee on Air and Water Pollution, hearings, *Air Pollution—1970, Part 1*, 91st Cong., 2d sess., 1970, p. 160.

20. Ibid., *Part 4*, p. 1489.

21. Ibid., p. 1492.

22. Ibid., p. 1498.

23. Ibid., p. 1498-1503.

24. Ibid., p. 1500.

25. John C. Esposito, *Vanishing Air* (New York: Grossman, 1970). See Jones, *Clean Air*, pp. 191-192, for a discussion of the book's impact.

26. U.S. Senate, Committee on Public Works, Report No. 90-403, *Air Quality Act of 1967*, 90th Cong., 1st sess., 1967, p. 28.

27. U.S. Senate, Committee on Public Works, "Committee Print No. 1," 91st Cong., 2d sess., 1970.

28. *National Journal* 1 (January 2, 1971):29.

29. U.S. Senate, Committee on Public Works, Report No. 91-1196, *National Air Quality Standards Act of 1970*, 91st Cong., 2d sess., 1970, p. 3.

30. Ibid., p. 10.

31. U.S. House of Representatives, Report No. 91-1783, *Clean Air Amendments of 1970*, 91st Cong., 2d sess., 1970, p. 5.

32. *Congressional Record* 116 (December 18, 1970):42381.

33. Jones, *Clean Air*, chap. 7.

34. John T. Middleton, acting commissioner, Air Pollution Control Office, EPA, "Briefing Memorandum: National Primary and Secondary Ambient Air Quality Standards" (n.d.).

35. *Federal Register* 36 (April 30, 1971):8186. Emphasis added.

36. U.S. House of Representatives, Committee on Interstate and Foreign Commerce, Subcommittee on Public Health and Environment, hearings, *Clean*

Air Act Oversight, 92d Cong., 1st and 2d sess., 1971-1972,p. 371. Emphasis added.

37. The company's comments are reproduced in U.S. Senate, Committee on Public Works, Subcommittee on Air and Water Pollution, hearings, *Implementation of the Clean Air Act Amendments of 1970–Part 2,* 92d Cong., 2d sess., 1972, pp. 491–493.

38. Houthakker's speech to the Cleveland Business Economists Club is reported in the *New York Times,* April 20, 1971. p. 47.

39. *Federal Register* 36 (April 30, 1971):8186.

40. U.S. Department of Interior, Bureau of Mines, *Minerals Yearbook, 1971* (1973), 1:941.

41. *Federal Register* 36 (April 7, 1971):6692.

42. In 1970, for example, the average sulfur content of coal used by electric utilities was 2.58 percent. See FPC, *Steam-Electric Plant Air and Water Quality Control Data, 1970* (1973), p. 1.

43. *Public Papers of the Presidents of the United States: Richard M. Nixon–1970* (1971), p. 1087.

44. *Environmental Health Letter* 10 (April 15, 1971):1.

45. John Quarles,. *Cleaning Up America* (Boston: Houghton Mifflin, 1976), p. 36.

6 Challenged Implementation

A government's regulatory activities have the potential to restrict an industry's operations and well-being. A natural tendency exists among affected industries to resist such interference by widely publicized criticisms of the regulatory agency or more discreet attempts to influence bureaucratic decision making. The latter course is especially appropriate when a regulatory body has reached a stage of administrative maturity or near debility. Such agencies tend to lack political support and have lost much of the stamina required to confront abuses of the public interest. In Marver Bernstein's classic analysis of regulatory agencies, he suggests that mature agencies are concerned primarily with maintaining harmonious relations and with avoiding conflicts with the regulated interests.[1] With mature agencies, regulated groups usually have little to fear. Regulations do exist, but they tend to be acceptable to the regulated interests, largely because they have been successful in co-opting or capturing the agency's leadership.

When an agency is new or relatively young, however, Bernstein observes a far different pattern of relations. New agencies tend to be ambitious, aggressive, and crusading. They take an expansive view of their role. Accordingly friendly relations with regulated groups are thought to be less important than curbing the abuses that led to regulation in the first place. From the perspective of regulated groups, however, they have "vital interests to protect against the onslaught of the regulators."[2] Bernstein continues: "While the enactment of a statute is clear proof that the about-to-be-regulated groups have lost the first round in the struggle against regulation, it is equally clear that they do not give up when the law is passed."[3]

If this assessment is correct, then a newly formed agency such as the EPA would likely be the object of a substantial amount of interest-group pressure. Moreover the analysis from chapter 1 suggests that regulatory guidelines are rarely final; they seem to find themselves permanently fixed on a government's agenda. This is especially likely for environmental standards that have been established on the basis of incomplete or poorly understood information.

A further reason to expect that the EPA would have problems after its creation stems from the fact that the agency's actions impinge on the prerogatives and territory of many of its senior colleagues, such as cabinet departments and regulatory commissions. Each of these agencies and departments represents a well-entrenched bureaucracy with relatively secure policy spaces. And like virtually all other agencies, each assumes that it has primary responsibility for the accomplishment of particular policy goals. To infringe on an agency's policy space is to evoke charges of interference and foul play. Much previous research supports this view and suggests that when an agency's jurisdiction is threatened, even marginally, its response is likely to be one of suspicion or what Anthony Downs has called "territorial vigilance."[4] The agency is likely to resist territorial encroachment and to appraise changes in policy "in the light of effects on its jurisdictional position rather than in the light of merit."[5]

In short, one can hypothesize that conflict among agencies is likely when their jurisdictions overlap. Furthermore the scope of this conflict is directly related to the number of overlapping policy responsibilities and to the extent or range of these overlaps. When conflicts do arise, they can be politically embarrassing, so it is to the advantage of public officials to mute these conflicts, perhaps by imposing or negotiating some form of cooperation or coordination among the competitors. As Richard Fenno has pointed out, however, "Before effective coordination can take place there must be an understanding of the goals toward which coordination will be pointed."[6] Hence if agencies with overlapping jurisdictions share common goals, the likelihood of coordination is enhanced; and if they seek disparate goals, the prospects for coordination are diminished. Despite this proposition, the difficulties of achieving coordination among like-minded organizations should not be underestimated. In a political system singularly reliant on incremental decision making, overall coordination of policy objectives is very difficult. Other factors that can reduce the possibility of meaningful coordination include jurisdictional pride, limited party responsibility, divided control of the executive and legislative branches, the lack of widely accepted and comprehensive goals, a lack of incentives for coordination, and a lack of collective responsibility among various agencies.

In the case of the EPA, the total coordination of its activities with those of other agencies may be impossible. Unlike most other government bureaus, the EPA was created to take a holistic perspective of a complex problem. In fulfilling its responsibility to halt deterioration of the environment, the EPA's activities in the areas of air, water, noise, pesticides, solid wastes, drinking water, and toxic substances collide with the programs of many other agencies. President Nixon offered a striking enunciation of this point when he proposed the creation of the EPA: "Almost every part of government is concerned with the environment in some way, and affects it in some way. Yet each department also has its own primary mission—such as resource development, transportation, health, defense, urban growth or agriculture—which necessarily affects its own view of environmental questions."[7] What Nixon failed to mention was that most departments ordinarily place their own missions ahead of environmental protection. This put the EPA in the unenviable position of trying to reverse this situation, and many areas of conflicting and overlapping jurisdiction have resulted.

Although overlapping jurisdictions are not unusual among bureaucratic agencies, the EPA is involved in more than its fair share. A study conducted by a House committee in 1975 underscores this point. As part of a larger study on federal regulatory agencies, the House Commerce Committee attempted to ascertain the actual or potential overlap of programs among nine regulatory agencies. In addition to the EPA, the study examined such well-known regulatory bodies as the Federal Communications Commission, the Federal Trade Commission, and the Securities and Exchange Commission. Each of these agencies has wide-ranging responsibilities, but the study found that these organizations are involved in relatively few jurisdictional overlaps. They range from the Interstate Commerce Commission's single overlap with the Federal Maritime Commission to the Securities and Exchange Commission's nine potential conflicts with such agencies as the Federal Power Commission and the Small Business Administration. Excluding the EPA, the average number of jurisdictional overlaps for the eight remaining agencies was slightly more than five. In contrast, the study found that the EPA was involved in at least forty-two.[8] In fact, the EPA's large number of overlaps puts the agency in the same league as many cabinet departments.

Unlike the overlap of other regulatory agencies, which frequently involve two agencies with compatible goals but different perspectives, many of the EPA's overlaps represent instances of incompatible goals. As an example, the EPA regulates pesticide products; the Department of Agriculture's goal is to increase the production of food and fiber. This mission has led Agriculture to encourage the use of certain pesticides, while EPA is trying to prohibit them. Many similar instances exist in which the EPA finds itself defending its actions against other government departments. Consequently the EPA must devote many of its limited resources to jurisdictional disputes instead of to efforts to accomplish its mandate.

It is clear that a seeming paradox exists. The EPA began its activities in 1970 with many reasons to believe that it would be successful in achieving acceptable levels of air quality. The analysis of regulatory policies and overlapping territories suggests, however, that the EPA's definition of clean air would be subject to continuous debate not only from industry but from other governmental units as well. In fact, this has been the case. Since the EPA first established national air quality standards for SO_2 in early 1971, these standards and the underlying support for them have been the subject of almost constant controversy. Some groups continue to argue that the standards are too lax, while others argue just the opposite.

The outcome of this debate is no less important today than it was in earlier years, and it may actually be more important. If the choice of standards for SO_2 has erred on the side of excessive stringency, then billions of dollars have been wasted on unnecessary control technologies, changes in industrial processes, and changes in the use and distribution of costly energy supplies. An error in the other direction means that the health of millions of Americans is endangered unnecessarily. In either case, incorrect standards for one pollutant bring into question not only the appropriateness of other air quality standards established by similar procedures but also the merits and credibility of all environmental programs.

These issues raise both scientific and political concerns, and it would be useful to separate them accordingly. Unfortunately such a division is neither feasible nor especially fruitful. Since 1971, scientific research and political decisions on SO_2 have been as inseparable as they were in the preceding years. Science has affected political decisions on the pollutant as much as politics has affected scientific activity, as a discussion on the Clean Air Amendment's implementation shows.

The First Challenges

With the passage of the 1970 legislation, the Congress excluded consideration of social and economic variables in meeting air quality standards. Earlier legislation had allowed and even encouraged such consideration, but the result had been procrastination and only negligible improvements in air quality. The Clean Air Amendments of 1970 were meant to change this. As Senator Muskie put it, air quality sufficient to protect the nation's health should not be delayed for reasons of cost or inconvenience. While this was Muskie's goal, he was not responsible for detailing how to achieve it. This job belonged to the EPA, but Muskie must have been heartened when William Ruckelshaus testified at his confirmation hearings: "I believe that regardless of the opinion of the administration or the opinion of the Administrator of this Agency, as to the wisdom or advisability of any section in the law, once the Congress passes the law and gives the responsibility to an administrative official to implement it or enforce it, his responsibility is clear and he must do what Congress has enacted into law."[9] This was an impressive statement since Ruckelshaus was promising his allegiance to a Democratic Congress. His mandate was tested much sooner than either Muskie or Ruckelshaus had believed possible. Within a few weeks of publication of the air quality standards and the proposed guidelines for state implementation plans, the EPA found itself embroiled in conflict over its scientific and administrative decision making.

Guidelines for State Implementation Plans

In light of the difficulties that the states had had in writing their implementation plans under the Air Quality Act, the EPA believed it was advisable to publish new guidelines consistent with the 1970 legislation. These guidelines would assist the states in developing a comprehensive strategy to achieve acceptable levels of air quality for all criteria pollutants. The EPA made its proposals in early April 1971, a few weeks before it published the new air quality standards. After an extended period for public comments on the proposed guidelines and EPA's consideration of the comments, the guidelines were readied for final publication. The revised guidelines were a pleasant surprise to many environmentalists. Many of their comments had led the EPA to make what they considered to be important improvements. These included provisions that required states to provide public access to emission data and clarified the need for states to de-velop compliance schedules for polluters. In contrast, the many comments from business and industry, including thirty from the electric utility industry, had very little effect; the guidelines said nothing about cost considerations and neglected many of industries' other social and economic concerns. There seemed to be general agreement that the proposed guidelines closely adhered to congres-sional intent, as one would have expected from Ruckelshaus's earlier comments.

The highest officials in the EPA reviewed the guidelines, approved them in June, and then recommended that Ruckelshaus promptly publish them. Instead he turned them over to the Office of Management and Budget (OMB) for its own review. The previous month, the OMB director, George P. Shultz, had told Ruckelshaus that special OMB review and clearance would precede formal publication of all EPA decisions affecting other federal agencies or imposing "significant costs on or negative benefits to nonfederal sectors." Shultz said that such a procedure was necessary because the EPA was providing "insufficient information on specific benefits of proposed environmental regulations."[10]

At the time of OMB's review, its officials were subject to neither senatorial confirmation nor automatic congressional oversight. The OMB was solely re-sponsible to the president, not to the Congress. President Nixon was well aware that a law's administrative implementation is of far greater consequence than is its legislative formulation. Thus the OMB provided Nixon with a useful means to scrutinize the EPA's performance. Indeed some evidence suggests that environ-mental activities ranked as a favorite target of OMB's oversight. During the OMB's review of the guidelines on implementation plans, one environmental newsletter went so far as to claim that Nixon used the OMB "as sort of a final filter for unwanted legislation," such as the Clean Air Amendments, which a Democratic Congress had given him.[11]

The OMB claimed that it was not conducting a review but "only giving other agencies a chance to comment on the guidelines' effect on their own areas of responsibility or their budgets."[12] At first glance, such an opportunity would seem unnecessary. When the EPA had originally proposed the guidelines in April, it had invited all interested parties, including all federal agencies, to submit written comments. Only five agencies had responded. Of these, only the response of the Federal Power Commission (FPC) had gone into much detail or expressed much opposition. The paucity of comments from within the federal government, however, did not deter the OMB from soliciting a second round of comments from the FPC and the Department of Commerce, two of the EPA's staunchest critics. Exactly what happened during the process is not entirely known, but by the time the OMB finished its review, the guidelines had been changed drastically.

As an illustration of the changes that had taken place, an appendix providing examples of administrative regulations that states might wish to adopt was especially hard hit. Sample regulations dealing with permit systems, emissions monitoring, and compliance schedules also had been deleted. Other changes "included a series of stipulations which, among other things, [were] intended to encourage States to consider the relative social and economic impacts of alternative control strategies before selecting the one(s) they [would] employ."[13] The regulations themselves encouraged the states to consider the costs and benefits of control measures and the social and economic impacts on energy, transportation, employment, and availability of fuels. These changes could serve only to bring about delay and compromise in achieving acceptable levels of air quality. Of similar importance was the fact that many of the changes had seemingly ignored congressional intent.

Although the final guidelines were published in August 1971, the members of Senator Muskie's environmental committee waited another six months before they questioned Ruckelshaus. When they did, he adamantly denied that the OMB had required any changes; instead he defended and assumed full responsibility for all the changes.[14] Ruckelshaus made a respectable effort to detail most of what had happened, but he failed to explain adequately why he had approved the original version of the guidelines published in April 1971 but rejected large parts of them in August. Similarly he did not clarify why he had overruled three of his top assistants, who had recommended approval of the earlier version. Finally he did not think it was inappropriate that the OMB review had taken place.

Whatever the explanation, the EPA's integrity as an independent and effective environmental advocate suffered in the eyes of the press, the public, and many congressmen. While it may never be possible to know the precise reasons for the change of heart, some speculation is in order. First, Ruckelshaus may have sincerely believed that the changes were necessary. He was trying to assume a practical approach in which air quality standards could be met in the most economical manner. No provision in the law required the use of the most expensive control technologies, and no provision required clean air to be achieved immediately. Ruckelshaus recognized that the guidelines would be a source of concern to industry, and there was no reason to alienate it by requiring unnecessary expenditures. The 1970 legislation had also maintained the primacy of state and local governments in the implementation phase, and too restrictive guidelines could usurp much of this primacy.

If Ruckelshaus was disenchanted with OMB's review, there is much he could have done to alert the Congress to the intrusion. It is not at all unusual, for example, for an agency head to notify friendly congressmen about interagency disputes, particularly those involving OMB. Another technique available to distraught agencies is to arrange for a sympathetic congressman to ask certain questions during hearings in order to bring out OMB interference.[15] The fact that Ruckelshaus did not use either of these ploys suggests his acceptance of OMB's involvement.

Second, Ruckelshaus was sufficiently astute to realize that the guidelines in any form would impinge on the activities and jurisdictions of several other federal agencies, such as the FPC and the Commerce Department. The mission of the FPC, which is now part of the Department of Energy, is to ensure an "abundant supply of electrical energy . . . with the greatest possible economy and with regard to the proper utilization and conservation of natural resources."[16] Thus the power commission was understandably skeptical about EPA's original guidelines, which suggested that many electric utilities should increase their use of natural gas at the expense of coal. The FPC claimed that if the EPA's recommendations were not amended, large-scale energy shortages would result. From the EPA's viewpoint, if modified guidelines could still lead to clean air, then it made no sense to ignore the FPC's position. In this instance, a change in the guidelines can be seen as practical deference to the FPC's expertise.

Third, partisan politics surely motivated OMB's interference. The Nixon administration was loathe to alienate the business community, especially since the president was gearing up for his reelection campaign. Even John Quarles, EPA's top legal officer and one of the administrator's closest aides, believed that OMB entered the scene for political reasons. In his own retrospective analysis, Quarles points out that OMB's involvement "reflected political pressures quietly exerting their force against the strong environmental requirements."[17] While Quarles admits that politics was involved, he does not develop his statement to indicate the nature, source, or strength of the political pressures.

Despite Quarles's omission, it is not hard to discern the source of the pressure. The Nixon administration was being responsive to industry's apprehension about governmental restraints. Many businesses and industries claimed that excessive regulation in the environmental area unfairly inhibited their growth and further development. One means of placating these critics would be to ensure that the OMB reviewed prospective regulations for their suitability before they were issued. In fact, the review of EPA's guidelines seemed to be so successful that the OMB soon initiated a procedure for interagency review of all "proposed agency regulations, standards, guidelines, and similar materials pertaining to environmental quality, consumer protection, and occupational and public health and safety."[18]

Fourth, Jones has suggested that the public support for environmental protection had begun to lag by 1971, and that the first indications of a public backlash had begun to develop.[19] The threat of plant closings, job losses, and higher prices due to pollution control equipment no doubt had an effect on public attitudes toward environmental protection. For an agency closely tied to and dependent on public support, the EPA could not afford to alienate its most important constituency and most important beneficiaries. By softening the guidelines, the EPA could defuse much opposition and could also stress that the changes represented a balanced and reasoned approach to environmental protection.

Nonsignificant Deterioration

Although the 1970 law required many heavily polluted areas to reduce substantially emissions of SO_2, still other parts of the country had very clean air that was far below the new standards. Many people who lived in these pristine areas were apprehensive that the standards would license massive increases in pollution. To prevent just such a situation, the Congress created the concept of nonsignificant deterioration (NSD), which precludes further contamination of ambient air that is below existing standards. Carried to its extreme, the application of NSD could mean an end to all economic and industrial growth in the cleanest areas of the country. The concept brought a new meaning to acceptability. On the one hand, acceptable air would be defined in terms of air quality standards. On the other hand, NSD meant that for large parts of the country, air quality could be unacceptable even if levels of SO_2 were below the standards.

Surprisingly the air pollution legislation had not specifically mentioned NSD. The Air Quality Act had simply stated that one of its purposes would be "to protect and enhance the quality of the Nation's air resources."[20] When the National Air Pollution Control Administration issued guidelines for the development of implementation plans in 1969, the agency had interpreted this phrase to mean that "air quality standards which, even if fully implemented, would result in significant deterioration in air quality . . . clearly would conflict with the expressed purpose of the law."[21] In making this assessment, there is no evidence to indicate that NAPCA had considered the policy's possible social or economic consequences. As in the past, the agency had probably been unprepared to do so. Because so little had happened as a result of the Air Quality Act, NSD had not become a major issue for NAPCA.

During congressional consideration of the Clean Air Amendments in 1969 and 1970, however, several spokesmen for the Nixon administration repeated NAPCA's position. Of course, it is likely that NAPCA had prepared this testimony, which simply concurred with its earlier position. In addition, the administration spokesmen were probably unaware of the wider implications of their stance. Regardless of their position, committee reports on the new law in both the House and the Senate agreed that protecting the nation's air quality meant preventing its deterioration.[22]

There was widespread support for NSD as a concept, but the EPA found itself in a familiar quandary when it tried to define the concept. When the EPA proposed the national air quality standards in January 1971, it noted that the "standards shall not be considered in any manner to allow significant deterioration of existing air quality in any portion of any State."[23] The EPA did not, however, indicate what constituted significant deterioration, and this absence is fairly easy to explain. Nowhere in the Clean Air Amendments or in their legislative history had there ever been an attempt to define the term or even to detail how such deterioration should be prevented.

If the concept was going to have any meaning, some definition would have to be provided. At first, the EPA's proposed guidelines for implementation plans had not contained a statement on NSD, but after public comment a statement was added that coincided with its January position on the air quality standards. When the final guidelines were published in August (after OMB's review), the statement had been changed. In what had been called a "sudden unilateral reversal" of congressional and administration policy, the guidelines indicated that deterioration could occur as long as secondary standards were not exceeded.

Ruckelshaus claimed that the guidelines had been changed because he did not believe the EPA had the legal authority to mandate NSD. There is no doubt, however, that the EPA had found itself in an awkward situation. The agency was surely in favor of clean air, and probably in favor of some form of NSD as well, but the hostile political repercussions of implementing a stringent regulation were more than the EPA could handle. President Nixon, who prided himself on his close ties to the business community, would not support a program that unduly curtailed economic development.

Although the final wording on NSD was acceptable to the Nixon administration, it was not satisfactory to the Sierra Club. Along with three other environmental groups, the Sierra Club filed a suit against Ruckelshaus in May 1972, charging him with a violation of congressional intent. The club was concerned that pollution below the secondary standards could be harmful. To support its position, the club relied on a statement from NAPCA that had concluded that no level exists below which air pollution is not harmful to humans.[24]

The Sierra Club had another concern that focused on the possible economic impacts if deterioration was not prevented. In most instances, metropolitan areas are more polluted than rural areas. In the absence of a strategy for NSD, clean rural areas and states could allow massive increases in pollution and thereby attract industries from the cities, where levels of pollution already approached or exceeded the air quality standards. Similarly industries interested in expansion could relocate in parts of the country where they could pollute without the need for expensive control technologies.

Less than a week after the Sierra Club had filed its suit and only a day before the EPA was required to accept or reject the states' implementation plans, the U.S. District Court in Washington, D.C., reached a decision. It agreed with the Sierra Club's position and prohibited the EPA from approving certain parts of the plans unless they contained adequate mechanisms to prevent deterioration. The next step was the EPA's, and whatever it chose would lead to trouble. To accept the decision without appeal would mean jurisdictional conflict with the OMB and other governmental agencies satisfied with the existing arrangements. To oppose the decision or to request clarification through an appeal would place the EPA in favor of more pollution. This situation appeared incomprehensible to many people who questioned Ruckelshaus the day after the court's decision.

Ruckelshaus insisted that he was not challenging the need to prevent deterioration, but within hours of the court's verdict, the EPA had already started work on an appeal. After considering the EPA's request to overturn the decision, the court of appeals refused to do so, unanimously affirming the position of the lower court.[25] Thus EPA would have to review all implementation plans in terms of degradation, even though the agency was still unsure of what the concept meant. Neither the district court nor the court of appeals had provided any guidance, and Ruckelshaus admitted that he was at a loss to explain NSD. Notwithstanding the lack of a definition, the EPA's review led it to reject parts of all plans in November 1972 because they did not have effective programs for NSD. Finally the EPA appealed to the U.S. Supreme Court, but it too refused to alter the lower courts' verdicts.[26]

The EPA again found itself in a predicament because it would have to decide what constituted adequate means to prevent deterioration. The agency eventually proposed four alternate plans in July 1973, with each one offering different methods to balance the concern for pristine air with social and economic considerations.[27] Each allowed for some deterioration because the EPA maintained that strict adherence to the concept would stop all economic growth. This position could lead only to further controversy. The Sierra Club had never opposed growth or even small increases in pollution, and it did not want to be miscast as a villain. At the same time, however, the club did not want its judicial victories to be hollow ones, but EPA's action seemed to ensure just such a situation.

Questions about Scientific Data

While there was much activity in the early 1970s about how the national air quality standards would be translated into meaningful action, there was also considerable discussion about the standards' scientific support. The Clean Air Amendments had included a provision that allowed judicial review of any primary or secondary standard as long as an appeal was initiated within thirty days of the standard's final promulgation.[28] Thus any appeal would have to be filed by the end of May 1971. Of the six criteria pollutants, only one was challenged. The Kennecott Copper Corporation filed an appeal on May 28 with the U.S. Court of Appeals in Washington, D.C., challenging one of the secondary standards for SO_2. Kennecott asserted that the standard limiting annual concentrations of SO_2 to 60 micrograms per cubic meter was not based on the criteria or adequately supported by other available data.[29]

The EPA did not lose the case; nor did it win. When the court issued its decision in February 1972, it requested the EPA administrator "to supply an implementing statement that will enlighten the court as to the basis on which he reached the 60 standard from the material in the Criteria."[30] To overcome the lack of documentation, the court specified that the administrator could revise the criteria on the basis of available data or, in their absence, revise the standard.

In response, the EPA summarized several studies said to corroborate the secondary standard that was under challenge. It also cited several studies that had found that concentrations of SO_2 as low as 40 micrograms per cubic meter over a period of several months harmed vegetation and suggested the possible need for a one-hour secondary standard, "in addition to the standards already promulgated."[31] To buttress its case further, the EPA made plans to publish new criteria for sulfur oxides. The Kennecott challenge involved only the secondary, nonhealth standards, but the agency decided to revise the entire criteria, including those sections on health-related effects.

To accomplish the revision, a special task force was established in May 1972. At this point, the details of what happened next are somewhat unclear. Nonetheless it seems that on June 1, the researchers in EPA's Division of Health Effects Research (DHER) were told that they would have only a month to draft four chapters for new criteria and to finish a monograph incorporating all of their current work on the health effects of sulfur oxides. After two months of working on the monograph and the draft criteria, a decision was made to split the two tasks. Work on the health-related monograph would continue, but further work on the new criteria was stopped, even though a draft of an entirely rewritten document had been finished in July. The EPA decided not to seek a review of the draft by its criteria advisory committee or by other federal agencies in accordance with the Clean Air Amendments. Several years later, when he was asked why the revised criteria had not been published, EPA's second administrator, Russell Train, indicated that he did not know.[32]

The Kennecott case involved only the secondary standards, so the EPA did seek review of the chapter in the criteria dealing with sulfur oxides and vegetation. This revised chapter was published in September 1973. At first, there was reason to believe that it might be used to justify more stringent secondary standards, or so it had seemed in August when a top EPA official announced that possible shortages of home heating oil would not lead to a relaxation of the standards.[33] In view of this statement and the agency's earlier claim that supportive data existed for the standard of 60 micrograms per cubic meter, what the EPA did next amounted to a complete turnabout. The agency left unchanged the secondary, three-hour standard of 1,300 micrograms per cubic meter, but it revoked both the annual secondary standard of 60 micrograms per cubic meter and the twenty-four-hour secondary standard of 260 micrograms per cubic meter. Although these changes did not affect the health-related standards, they did bring into question the EPA's credibility and objectivity in using scientific data.

Further Scientific Research

The EPA's Health Effects Research Laboratory had been given a dual task in May 1972: to publish new criteria and to write a health-effects monograph. This second task referred to work on the Community Health and Environmental Surveillance System (CHESS). The CHESS studies represent the most ambitious analysis of air pollution ever undertaken in the United States. One of their primary purposes was to provide "a sound basis for relating human health effects to precise quantities of SO_2 and other oxides of sulfur in air and to provide a reliable record of the improvement of the health of the U.S. population as air pollution was controlled and reduced.[34] The CHESS researchers attempted to use classic research designs in a long-term epidemiological study. They selected sets of communities within larger jurisdictions on the basis of differing exposure levels to air pollutants and matched them as closely as possible on the basis of their demographic, socioeconomic, and climatological characteristics.

The study had the potential to remedy many deficiencies in existing knowledge and to dispel most objections to the air quality standards. If it could be completed successfully, the EPA would be in possession of incomparable data. The appropriateness of the air quality standards could be evaluated, margins of safety could be determined with greater precision, the possible need for new standards could be ascertained, and the health benefits of air pollution control could be pinpointed with improved accuracy. The research would not be easy. Successful completion would require the coordination of hundreds of researchers and technicians, as well as the constant cooperation of tens of thousands of carefully selected respondents who would be asked to maintain a detailed record of any changes in their health. The project would cost millions of dollars and collect millions of bits of information. Under the best of circumstances, the possibility of error at every stage of the study was great.

In the early parts of the project, most of the effort was directed at the health effects of sulfur oxides, but the researchers also examined the related role of sulfates and particulates. The initial monograph, which was to contain approximately eighteen separate studies, was intended to examine and report on data collected on these pollutants during the project's first year (1970-1971). To ensure consistency and the elimination of repetitive material in the separate studies, John F. Finklea, then the director of the HERL, read each manuscript and assumed responsibility for requiring recomputation of some statistical analyses and for resolving differences in interpretation among the many authors.

Once work had ceased on the revised criteria, Finklea and his colleagues were able to devote their full attention to the CHESS program. Nonetheless they still had little time for pause or reflection. EPA's headquarters was anxious for results and wanted a monograph ready for external review as quickly as possible. The time available was indeed short, as one participant later recounted: "If you remember, [one researcher] . . . commented that he would like about 1 to 2 years to analyze these studies. And most of the analyses that went into this [CHESS] document were done in 3 weeks."[35] The monograph was put together hurriedly. A draft ready for internal review was completed by August 1. After Finklea's review, a draft was released for external review in October 1972.

In the belief that outside reviews were essential, efforts were made to obtain comments from a wide variety of sources, including several trade and industrial groups. In all, comments were solicited from nearly 150 outside reviewers. By the new year, many of the reviewers had responded, and it was now possible to consider and respond to the critical evaluations.

The final monograph would not be published until 1974, but its findings were already under wide discussion in early 1973. After the *Washington Post* obtained a preliminary draft, the newspaper's headlines read: "Study Finds Pollution Laws Too Lax to Protect Health."[36] The study, said the *Post*, had found that existing standards for sulfur oxides might not be strict enough to protect the public's health. While the annual standards for SO_2 and particulates were not questioned, the CHESS researchers had supposedly found that adverse effects can be detected even though the twenty-four-hour standards for the two pollutants are not violated. In some situations in which the short-term standards are achieved, the study noted that SO_2 and particulates interact in a poorly understood way to produce sulfates, which include sulfuric acid, sulfites, and neutral metallic sulfates and can aggravate asthma and the heart and lung ailments of some elderly people.[37] Instead of blaming the aggravation on SO_2 and particulates, the researchers emphasized that adverse effects should probably be attributed to sulfates in the atmosphere. In other words, the report suggested that the by-products of SO_2 might be more harmful than the single pollutant by itself.

The incrimination of sulfates, while not new, had tremendous implications. From a health perspective, if sulfates are guilty, separate standards for SO_2 and particulates are not entirely appropriate. The finding also meant that it might not make sense to measure short-term ambient levels of SO_2 since they are probably poor indicators of likely adverse effects.

From a control perspective, the possible ramifications were equally important. If short-term levels of SO_2 are not solely to blame for health damage, then combinations of pollutants acting together must be considered in developing control strategies. To do otherwise would mean that a polluter in compliance with air pollution regulations could still contribute to unacceptable levels of air quality. Polluters would have sufficient cause to complain about inappropriate regulations, and regulators would be hardpressed to justify standards that did not ensure protection of the public's health. These were significant implications, but the CHESS researchers offered little guidance for policy makers. First, the researchers emphasized that any measures increasing the amount of sulfur in the air should be avoided. Coincidentally this recommendation came at about the same time that EPA officials were proposing the opposite tactic for nonsignificant deterioration. Second, the researchers stressed the point that sulfates should not be ignored but then concluded that not enough was known about them to set an air quality standard.

Although the CHESS studies had several scientific purposes, they were also intended as a means of revalidating the EPA's original value judgment in its choice of standards for SO_2. Because the findings were said to justify these standards, the EPA made frequent use of the CHESS findings well before they were formally published. In less than a month's time in the fall of 1973, for example, the EPA made extensive use of the data in testimony before two congressional committees and a special conference on the health effects of air pollutants that the National Academy of Sciences sponsored. At one of the congressional hearings, John Quarles, who had become the second-ranking official in the EPA, said that the agency's research suggested that it might be necessary to tighten the air quality standards in the future.[38] A few days later, Russell Train, the EPA's new administrator, told another committee of his concern for an accurate SO_2 standard:

> The sulfur standard has enormous consequences with regard to the energy picture, and an improperly set standard could impose unwarranted social costs. . . . With respect to the sulfur dioxide standard, all subsequent research has only served to confirm our original . . . standards. A series of recent epidemiological efforts known as the CHESS studies has not only served to reestablish the validity of the original standard, but has highlighted the possible hazards posed by sulfates.[39]

Despite Train's apparent certainty, the CHESS researchers themselves were not convinced that the standards were adequate or even that enough information was available to make such a judgment. As Finklea pointed out, "We lack an adequate information base to assure scientifically incontrovertible primary ambient air quality standards."[40] He said that an adequate number of epidemiological studies existed but noted that supporting data for the SO_2 and particulate standards were sorely deficient in the clinical and toxicological areas (table 6-1). The paucity of supporting studies noted in table 6-1 illustrates the claim that the air quality standards rested on weak scientific footing. In some cases, the absence of adequate data means that it is impossible to know if the SO_2 standards offer any protection from certain kinds of effects. For some other effects, the absence of clinical and toxicological studies means that the standards

Table 6-1
Adverse Health Effects That Might Be Attributed to Exposures Involving Sulfur Oxides and Suspended Particulates

Expected Effect	Epidemiology	Research Approach Clinical Studies	Toxicology
Increased susceptibility to acute respiratory disease	Multiple studies	No data	Isolated studies
Aggravation of asthma	Few replicated studies	No data	No data
Aggravation of heart or lung disease	Multiple studies	No data	No data
Irritation symptoms	Multiple studies	Few replicated studies	No data
Altered lung function	Multiple studies	Few replicated studies	Multiple studies
Increased risk of chronic lung disease	Multiple studies	No data	No data
Cancer	No data	No data	No data
Congenital defects	No data	No data	No data
Impaired defense mechanisms	No data	No data	No data

Source: U.S. House of Representatives, Committee on Science and Astronautics, Subcommittee on Energy, hearings, *Energy and Environmental Standards*, 93d Cong., 1st and 2d sess., 1973, p. 474.

rest solely on epidemiological studies, which are not entirely appropriate for setting air quality standards.

For the epidemiological studies that were available, however, Finklea asserted that levels of SO_2 below the twenty-four-hour primary standard could increase the likelihood of premature death among some groups in urban areas. The short-term standard is 365 micrograms per cubic meter, but the point at which premature deaths occurred could be as low as 30 micrograms per cubic meter over a twenty-four-hour period, said Finklea.[41] In other instances, he repeated the claim that other adverse effects are linked more to sulfates than to SO_2 acting alone. Finklea also addressed the adequacy of the existing margins of safety, and he found these to be modest (for SO_2) and completely absent for sulfates, for which no standard exists.[42]

In sum, the CHESS studies provided the EPA with data of far-reaching importance. The studies had done more than just support the standards; they had presented the EPA with information that could not be ignored. Some segments of the public have their health jeopardized when short-term standards are achieved, and many more people suffer because of sulfates. What the agency would do with this information was not immediately apparent, but it was clear that the answer would be controversial in the face of growing opposition to environmental regulations.

Notes

1. Marver H. Bernstein, *Regulating Business by Independent Commission* (Princeton: Princeton University Press, 1955), p. 88.

2. Ibid., p. 79.

3. Ibid., p. 82.

4. Anthony Downs, *Inside Bureaucracy* (Boston: Little, Brown, 1966), p. 215.

5. Arthur Maass, *Muddy Waters* (Cambridge: Harvard University Press, 1951), p. 253.

6. Richard F. Fenno, Jr., *The President's Cabinet* (New York: Vintage Books, 1959), p. 142.

7. The President's Message on Reorganization Plan No. 3 of 1970 is reprinted in U.S. House of Representatives, House Document 366, 91st Cong., 2d sess., 1970.

8. U.S. House of Representatives, Committee on Interstate and Foreign Commerce, Subcommittee on Oversight and Investigations, Subcommittee Print, *Federal Regulation and Regulatory Reform*, 94th Cong., 2d sess., 1976, p. 501.

9. U.S. Senate, Committee on Public Works, hearings, *Nomination of William D. Ruckelshaus,* 91st Cong., 2d sess., 1970, p. 32.

10. Cited by National Research Council, *Decision Making in the Environmental Protection Agency* (Washington, D.C.: NAS, 1977), p. 86.

11. *Air and Water News* 5 (August 2, 1971): 1.

12. Ibid. (July 12, 1971): 1.

13. EPA, Office of Air Programs, "Briefing Memorandum No. 2: EPA Regulations for Preparation of State Implementation Plans" (August 3, 1971).

14. U.S. Senate, Committee on Public Works, Subcommittee on Air and Water Pollution, hearings, *Implementation of the Clean Air Act Amendments of 1970–Part 1 (Title I)*, 92d Cong., 2d sess., 1972, p. 243.

15. Ruth C. Silva et al., *American Government* (New York: Alfred A. Knopf, 1976), p. 348.

16. The Federal Power Act, 16 USC 824a(a), sec. 202a.

17. John Quarles, *Cleaning Up America* (Boston: Houghton Mifflin, 1976), p. 88.

18. Memorandum from OMB Director George P. Shultz to Heads of Departments and Agencies, October 5, 1971.

19. Charles O. Jones, *Clean Air* (Pittsburgh: University of Pittsburgh Press, 1975), pp. 248–249.

20. PL 90-148, sec. 101(b) (1).

21. HEW, NAPCA, "Guidelines for the Development of Air Quality Standards and Implementation Plans" (1969), sec. 1.51.

22. U.S. House of Representatives, Committee on Interstate and Foreign Commerce, Report No. 91-1146, *Clean Air Amendments of 1970*, 91st Cong., 2d sess., 1970, p. 5, and U.S. Senate, Committee on Public Works, Report No. 91-1196, *National Air Quality Standards Act of 1970*, 91st Cong., 2d sess., 1970, p. 11.

23. *Federal Register* 36 (January 30, 1971): 1503.

24. HEW, "Guidelines," sec. 2.10.

25. *Sierra Club* v. *Ruckelshaus*, 344 F.Supp. 253 (D.D.C. 1972), *affirmed.* 4 E.R.C. 1815 (D.C. Cir. 1972).

26. *Fri* v. *Sierra Club*, 412 U.S. 541 (1973).

27. For a discussion of the plans, see Bruce M. Kramer, "Economics, Technology, and the Clean Air Amendments of 1970: The First Six Years," *Ecology Law Quarterly* 6 (1976): 224–228.

28. PL 91-604, sec. 307(b)(1).

29. *Kennecott Copper Corp.* v. *EPA*, 3 E.R.C. 1682 (1972).

30. Ibid., at 1685.

31. *Federal Register* 37 (May 12, 1972): 9578.

32. U.S. House of Representatives, Committee on Science and Technology, Subcommittee on the Environment and the Atmosphere and Committee on Interstate and Foreign Commerce, Subcommittee on Health and the Environment, joint hearings, *The Conduct of the EPA's "Community Health and Environmental Surveillance System" (CHESS) Studies*, 94th Cong., 2d sess., 1976, p. 89.

33. *Air/Water Pollution Report* 11 (August 27, 1973): 343.

34. U.S. House of Representatives, Committee on Science and Technology, Subcommittees on Special Studies, Investigations and Oversight and the Environment and the Atmosphere, Committee Print, *The Environmental Protection Agency's Research Program*, 94th Cong., 2d sess., 1976, p. 11.

35. House, *Conduct of the EPA's Studies*, p. 77.

36. *Washington Post*, February 12, 1973, p. A9.

37. EPA, Office of Research and Development, *Health Consequences of Sulfur Oxides: A Report from CHESS, 1970-1971* (Research Triangle Park, N.C.: National Environmental Research Center, 1974), p. 7-20.

38. U.S. House of Representatives, Committee on Interstate and Foreign Commerce, Subcommittee on Public Health and the Environment, hearings, *Clean Air Act Oversight—1973*, 93d Cong., 1st sess., 1973, p. 92.

39. U.S. House of Representatives, Committee on Science and Astronautics, Subcommittee on Energy, hearings, *Energy and Environmental Standards*, 93d Cong., 1st and 2d sess., 1973, p. 452.

40. House, *Clean Air Act Oversight—1973*, p. 79.

41. Ibid., p. 57.

42. Ibid., p. 78.

7

Second Thoughts on the Desirability of Clean Air

Anthony Downs has suggested that many problems of crucial importance suddenly leap into prominence but then gradually fade from attention as newer problems arise and as the public begins to appreciate the costs of solving the initial problems.[1] Writing in mid-1972, Downs placed public concern for environmental protection in the midst of just such a decline.

Some evidence can be marshaled to suggest that Downs's prediction had some merit as far as the public's commitment to clean air. Until 1972, the Clean Air Amendments had little impact on the public because most of the time immediately after their passage had been devoted to such procedural matters as publishing air quality standards and guidelines for state plans. Thus a national framework for enforcement was not established until May 1972, when the EPA granted partial approval to most implementation plans. After these approvals, the major social and economic costs of cleaner air became apparent. Downs's work suggests that the costs of achieving cleaner air would bring about a loss of public concern. In the present context, the question is whether such a shift in public opinion occurred and, if it did, how it affected the attitudes of policy makers toward the definition of acceptable levels of SO_2. Before looking at these issues, however, it is useful to examine the responses of the coal and electric utility industries to the stationary-source provisions of the Clean Air Amendments. Restrictions on emissions of SO_2 affect these industries more than any others, so they would have the most to gain by delayed implementation or by the EPA's loss of support.

Problems with Industry

The passage of the 1970 law did not signal the acquiescence of the coal and electric utility industries. Although they had lost many major battles in the fight over the specific provisions of the law, this did not ensure their defeat in subsequent battles over the law's implementation. Many parts of the law were imprecise, and the EPA would have to make many decisions about how the law should be interpreted and implemented. In its deliberations, the EPA could not safely ignore the wishes of the coal and electric utility industries. Without their cooperation, efforts to achieve acceptable levels of SO_2 would be impossible, so the industries possessed great potential leverage.

Amiable negotiations between the industries and the EPA appeared unlikely, however, because of the agency's relative youth and its crusading spirit,
characteristic of most new regulatory agencies. If satisfactory negotiations
seemed improbable, the industries could rely on traditional forms of resistance
to regulation. They could complain loudly about the inappropriateness of EPA's
decisions and bring their case to the attention of the Congress and the president.
They could also mobilize support from other governmental agencies friendly to
their interests. If successful, these tactics would involve the EPA in jurisdictional
battles that would surely require congressional and presidential involvement. For
a mature regulatory agency, the tactics would ordinarily be sufficient to demonstrate the industries' strength and to ensure the agency's willingness to compromise. In contrast, a young agency such as the EPA would be less amenable to
outside pressures and would be likely to respond in an aggressive manner.

An examination of EPA's choice of control strategies provides a good opportunity to examine the validity of these assumptions about the relations between the EPA and the two regulated interests. In order to reduce emissions of
SO_2, the EPA had originally believed that new power plants could be built to
burn natural gas or low-sulfur oil or that existing plants could switch from coal
to these other fuels. Many utilities began to use less polluting fuels, but it soon
became evident that the EPA had overestimated their widespread availability.

In the absence of sufficient quantities of acceptable fuels, an alternative is
to install pollution control equipment. One such technology relies on a technique that chemically removes SO_2 after combustion in boilers but before emission into the ambient air. These so-called scrubber systems are appealing because
they allow power plants to use most high-sulfur coals anywhere in the country,
including the most heavily polluted areas. The systems obviously enhance coal's
appeal as a source of energy.

Scrubbers were not commercially available in the late 1960s, so the National
Air Pollution Control Administration had encouraged utilities to use other ways
to reduce SO_2, including tall stacks, which disperse but do not reduce pollution,
and intermittent control systems (ICS), which rely on predictions of meteorological conditions and short-term changes in fuel use to achieve air quality standards. Tall stacks, some of which are over a thousand feet high, are especially
preferable to power companies because the stacks involve little technological
innovation and are relatively inexpensive to construct and operate.

Intermittent systems frequently include tall stacks as part of an overall
strategy. These systems usually require some costs for operation because companies relying on this scheme must have up-to-date weather information, as well
as the ability to shift to low-sulfur coals quickly if conditions indicate the need
to do so. In the late 1960s and early 1970s, many companies viewed the tall
stack-ICS option as an effective and reasonably priced means of achieving
acceptable levels of air quality. When properly used, many experts agree that

both methods ensure that air quality standards will be met in areas immediately adjacent to the source of pollution (but not necessarily in areas farther away).

NAPCA had approved the use of tall stacks and ICS, but this consent vanished soon after the EPA's creation. One section of the guidelines for implementation plans proposed in early 1971 stated that scrubbers were available for coal-burning power plants.[2] In contrast, tall stacks and ICS were not included because the EPA did not believe them to be sufficiently reliable. The agency initially took the position that the methods did not prevent violations of air quality standards and that higher levels of pollution in the atmosphere were undesirable.

After its announcement that scrubbers were available, the EPA vigorously prompted their widespread use, especially since the federal government had subsidized much of their developmental costs. Nonetheless many coal-burning utilities remained unconvinced of the need for new controls or, for that matter, the need for any changes in their environmental activities. Some claimed that scrubbers were unnecessary in view of such alternative strategies as tall stacks and because of the scrubbers' high costs, unavailability, and alleged unreliability. The utilities' opposition to scrubbers took many forms, most of which were bitterly critical of the EPA. Such reactions were not unexpected. The EPA's position on scrubbers represented a major change. NAPCA had encouraged utilities to reduce their pollution by whatever means they chose. The EPA altered this arrangement and tried to tell the utilities exactly what kinds of control strategies would be acceptable. Many utilities resented the EPA's intrusion into what they considered to be their own business.

The utilities were not alone in their opposition to the EPA's position on scrubbers. At various times in the early 1970s, the CEQ, the FPC, the Department of Interior, the Tennessee Valley Authority, and the President's Office of Science and Technology all sided with the utilities and opposed the EPA. These jurisdictional squabbles encouraged the utilities to resist the EPA's pressure to install scrubbers and also made the EPA's task more difficult in the absence of cooperation from other federal agencies.

The EPA did not take these criticisms lightly. Many times during the early 1970s, it accused the electric utilities of recalcitrance and a lack of social responsibility. In an unprecedented series of public hearings in 1973, for example, the EPA reviewed the status of utility compliance with regulations limiting the emissions of SO_2. The summary of the hearings roundly condemned the industry and said that the "testimony of some utilities indicates that they have applied greater efforts to defending their lack of progress or to attempting to change existing emission requirements than they have in controlling their SO_x emission through [scrubber] technology."[3]

The EPA found itself opposed by the coal industry as well. This industry had one major concern: ensuring that its major customers did not curtail their use of coal. Consequently the industry favored policies that encouraged consumption and opposed those that had the opposite effect. This strategy meant that the well-being of the coal industry was inseparable from that of its largest single customer, the electric utility industry. In the face of electric companies' inability or unwillingness to install control technologies that allow the use of high-sulfur coal, the coal industry found that it was losing much of its largest market as utilities switched to oil, natural gas, and nuclear power. The Clean Air Amendments' requirement that states include emission standards in their implementation plans further accelerated the trend away from coal. A representative from the coal industry cited the situation in Illinois as an example. The state's entire reserves of bituminous coal, the largest in the country, could not be used anywhere in the state because of restrictions on its allowable sulfur content.

Considering the impact of all implementation plans, the CEQ once estimated that they would eventually prevent the annual use of nearly 155 million tons of coal. To compensate for the precluded coal, the CEQ predicted that 584 million barrels of low-sulfur oil would have to be imported.[4]

With such dire predictions about the future of coal in the United States, it was not surprising that the National Coal Association was a major critic of environmental legislation. Carl Bagge, the president of the association, complained that the desire for clean air was overriding all other national goals. "The resultant side effects," said Bagge, "include a dangerous and addictive dependence on foreign fuels with its grave consequences for national security and the conduct of our foreign relations."[5] Bagge was not entirely correct in blaming the decline in coal usage on air pollution laws, but general agreement does exist that environmental regulations were the most important cause.[6] The decline in the use of coal and the related increase in the use of oil to produce electricity was especially noticeable in the Northeast. In Massachusetts, for example, coal had been used to produce over 36 percent of the state's electricity in 1967; this figure had declined to less than 1 percent in 1972. Over the same period, the contribution of oil to electrical production in the state rose from 52 percent to slightly over 90 percent.[7]

In addition to blaming air quality laws for declining coal usage, Bagge also claimed that state and federal action on the environmental front had cost his industry thousands of jobs and threatened many segments of it with extinction. It would have been impractical and self-serving for the coal interests to criticize the goal of improved environmental quality, but this did not deter them from making suggestions on how that goal ought to be achieved, as this dialogue with Senator Muskie in 1973 indicates:

Senator Muskie: I understand your objective. But the means, am I accurate in describing the means that you propose, that is repeated extensions of deadlines for health standards, is that correct?

Mr. Bagge: That is right.[8]

Although the coal and electric industries wanted major changes in the implementation of the environmental legislation, these groups would not be the sole determinant of change. For the most part, the extent of their success would depend on the amount of support that the EPA could generate and maintain from other sources. With strong support from the public, the Congress, and the president, the EPA could withstand any assault from regulated groups. In the absence of such support, however, the regulated groups would enjoy their greatest success.

Public Opinion and the Environment

At the beginning of the 1970s, a nationwide sample of Americans had been asked to list the two or three top problems facing them. Slightly more than 40 percent of the respondents identified air and water pollution as one of their major concerns. The respondents ranked pollution second among all items; others they identified included unemployment (35 percent), the war in Vietnam (31 percent), taxes and spending (31 percent), crime (28 percent), and drugs (18 percent). When the Harris Survey asked a nearly identical question in 1973, pollution had fallen to sixth place in the overall rating scheme; only 11 percent of the respondents listed it as one of their major concerns.[9]

These figures reflect a decline in concern for environmental protection relative to other national problems but not necessarily the loss of interest that Downs had suggested. In fact, data are available to refute Downs's dire predictions about the fate of the environmental movement. For example, when a representative sample of citizens was asked how they felt about reducing air pollution in 1972, 88 percent said that they were worried or concerned at least a great deal or a fair amount. Two years later, when the same question was asked of another group of Americans, the decline in concern was only six percentage points. This slight decline can be explained by the fact that a majority of Americans believed that at least some progress had been made in the fight against air pollution between 1972 and 1974.[10] In other words, because many people perceived air pollution to be a less serious problem in 1974 than in 1972, the need for public concern had diminished. This logic provides a partial explanation for the small change in public attitudes about air pollution. At the same time, however, the logic ignores the impact of a highly salient sequence of events that one would have expected to devastate public support for environmental protection.

During the winter of 1972–1973, Americans faced a severe shortage of energy supplies. Factories and schools were closed, and heating oil was rationed in many eastern states. By the fall of 1973, the situation had deteriorated even more. The Arab-Israeli war provoked many Arab countries to impose an embargo on all shipments of oil to the United States. The result was near chaos.

People waited hours to buy gasoline for their cars, and many observers predicted a long period of economic stagnation and high unemployment.

When the embargo was lifted in January 1974, Arab oil was back on the market but at sharply higher prices. The members of the Organization of Petroleum Exporting Countries had more than quadrupled the price of a barrel of crude oil in less than six months. The price increases dramatically reduced consumers' purchasing power and especially affected people on fixed incomes such as the elderly. Pennsylvania's Governor Milton Shapp gave a graphic description of the problem: "It is inhuman and malignant for a society in this wealthy nation to tolerate the specter of old people eating dog food as a trade-off for heat, light or rent money."[11]

If anything was ever going to threaten public support for cleaner air, the energy crisis seemed to be the prime candidate. Yet very few members of the general public preferred that environmental regulations be relaxed or deferred as a means of responding to the energy shortages. One explanation for this finding is that very few Americans believed that such regulations had caused the shortages. Moreover the shortages and the inconveniences caused by them did not increase the number of Americans who believed the national government was spending too much on environmental protection. In the spring of 1973 and a year later, after the oil embargo, for example, only about 8 percent of the public wanted the government to cut spending on environmental protection in contrast to the more than 60 percent in both years who wanted to see more spending.[12] These figures provide compelling evidence that the public had not become disenchanted with the environmental movement because of the energy situation.

The public's perceptions of the energy crisis were at great variance with those of some key policy makers and many businessmen. Even some of the EPA's top officials feared that its programs were on the point of collapse because of a loss of public support. John Quarles, who was now the EPA's deputy administrator, declared that the "energy crisis plunged the environmental movement into its most desperate struggle for public support. . . . A major rollback of environmental progress seemed imminent."[13] Many businessmen agreed with Quarles's prediction and seized the opportunity to accuse environmentalists of virtually every evil. These accusations included claims that environmental laws and their supporters had prevented construction of the Alaskan pipeline and several domestic oil refineries, had reduced the availability of domestic coal by imposing restrictions on strip-mining operations and limitations on the sulfur content of most coals, had increased the demand for certain fuels by requiring the installation of energy-intensive pollution control equipment, and had imposed unnecessarily rigid compliance dates to meet air quality standards.

The solution to the nation's energy problems, many businesses emphasized, would be to eliminate many environmental constraints or at least to delay their implementation. The SO_2 standards became a major target because they were believed to be crucial to a solution to the energy situation. Virtually all major users of coal and oil expressed their belief that relaxation of environmental standards was a necessity if the country wanted to expand its available energy supplies. Bagge of the National Coal Association argued that air pollution standards were the most important factor inhibiting greater use of coal. His solution to the energy problem was simple—relax or delay the implementation of air quality standards for SO_2 and then require as many power plants as possible to switch to coal and natural gas.[14] Many congressmen seemed receptive to just such a plan, and legislation was introduced that would have authorized the president to require power plants and industrial facilities that burned petroleum or natural gas to convert their boilers so that they could use coal on a permanent basis. These demands put the EPA on the defensive, and only one question remained: How would the president and the Congress respond to the conflicting viewpoints of the public and industry?

Energy and the Nixon Administration

On June 4, 1971, President Nixon delivered the nation's first presidential message on energy to the Congress. In his address, he listed a series of programs designed "to ensure an adequate supply of clean energy for the years ahead." While recognizing that the quest for clean air and increased supplies of energy could be in conflict, there was no reason they had to be. As Nixon explained it, Americans could "continue to know the blessings of both a high-energy civilization and a beautiful and healthy environment."[15] Almost two years later, in a message to the Congress on natural resources and the environment, Nixon repeated his belief that environmental quality did not have to be a victim of energy development.[16]

Although Nixon publicly stressed compatibility, many of his administration's actions belied this theme. Before the president left office, his administration initiated many actions that cast serious doubt on White House support for efforts to abate air pollution. As one illustration, the OMB asked HEW to conduct a study on the health effects of SO_2. This request came in the form of a letter in June 1973 from Roy Ash, OMB director, to Secretary of HEW Caspar Weinberger asking the latter's agency to "take the lead in a cooperative study with the EPA to document and examine the adequacy of available scientific information on the relation between sulfur oxides and their effects on human health."[17] The study would not be an effort to collect new information because Ash was interested only in a summary and evaluation of existing knowledge. The OMB wanted such an analysis, it was reported, because the Senate Subcommittee on Air and Water Pollution would soon begin hearings on possible revisions to the Clean Air Amendments, and OMB was concerned about the law's economic impacts. According to one source, OMB felt that it was imperative to explain the benefits of controlling SO_2 in health and other areas because it was interested in allowing states to set their own secondary standards by repealing the federal ones.[18]

Although requests for informational and evaluative reports are common in Washington, OMB's request was noteworthy because the EPA was not given primary responsibility for the task. The EPA had established the SO_2 standards on the basis of the best available scientific information in 1971. The following year the agency had prepared (but not published) updated criteria on sulfur oxides, and, at the time of Ash's request, preliminary drafts of the CHESS studies were available. Thus it was likely that the EPA could have provided Ash with whatever documentation he wanted in a relatively short time.

The EPA was also in a position to provide Ash with information about the economic impacts of the clean air legislation. The Air Quality Act of 1967 had required the secretary of HEW to conduct "a comprehensive study of the economic impact of air quality standards on the Nation's industries, communities, and other contributing sources of pollution, including an analysis of the national requirements for and the cost of controlling emissions to attain such standards of air quality."[19] The 1970 Clean Air Amendments had repeated this requirement, and the EPA had already completed several supplemental reports on economic impacts. In this respect, the EPA had both the experience and the expertise to detail the economic costs and benefits of pollution control.

These arguments apparently did not affect Ash, and the task of fulfilling his request was given to David Rall, director of HEW's National Institute of Environmental Health Sciences (NIEHS). This institute devotes most of its efforts to biomedical research on the relation between man's health and chemical, physical, and biological agents in the environment. Thus although selection of the NIEHS was jurisdictionally inappropriate because EPA had major responsibility for air pollution research, the choice of NIEHS was not without some merit. Rall and his staff were well qualified to evaluate existing research on sulfur oxides, though they were probably unqualified to consider the economic implications of air pollution. This latter task would surely be conducted inside the OMB.

Rall finished his work in early October 1973. The report, "A Review of the Health Effects of Sulfur Oxides," noted that the EPA's current primary standards for sulfur dioxide were appropriate and need not be changed.[20] Due to a lack of conclusive information, however, the report maintained that "further scientific information will be required to either validate the present standards or to justify alterations to these standards." This was a conclusion with which the EPA could easily agree. The EPA had never claimed that its information was complete, and its own continuing research on sulfur oxides reflected this.

Perhaps the most important findings of the Rall report were twofold. First, primary standards for SO_2 might not be perfect, but at least they appeared to be satisfactory in view of existing knowledge. Second, sulfates were probably more harmful than SO_2 or particulates by themselves. One conclusion with which Rall and the CHESS researchers agreed was that adequate margins of safety might not exist under any circumstances when SO_2 was converted into particulate sulfates.

In any case, Rall felt that existing knowledge on sulfur oxides was inadequate. A remedy was crucial since a decision based on faulty data could lead to one of two unfortunate consequences: "(1) accepting the lowest levels (worst case) to be significant to health when this is not true and consequently wrongly promoting expensive pollution controls, or (2) accepting the highest levels (least case) to be significant when this is not true and consequently wrongly allowing many people to suffer the ill effects of pollution toxicity."[21] Such uncertainty, said Rall, could lead to disaster for future generations. To overcome the deficiencies in supporting data, the report recommended further research on SO_2 and sulfates that would cost at least $30 million before significant results could be obtained.

It is unclear whether Rall's report provided Ash with the information he wanted. At the least, the report certainly did not provide any justification for a relaxation of SO_2 standards. Indeed Rall's concern for the sulfate problem suggested that it would be inappropriate to tamper with the standards as a means of combating the energy crisis. Ash's response to the report was probably indicative of his feelings toward it. Shortly after its release, the OMB announced a $5 million budget increase for research directed at the health effects of sulfates. Once this announcement was made, the Nixon administration made no further references to the report in congressional testimony on means of alleviating the energy problem or on ways of amending the clean air legislation.

Proposals for Change

When Ash requested the NIEHS to summarize knowledge on sulfur oxides, few people knew how he intended to use the results. Senator Muskie, however, believed the study was intended to justify a weakening of efforts to abate SO_2, but he was wrong in one important respect.[22] The OMB did not wait for the Rall report before initiating changes in the fight against the standards.

After claiming for months that the temporary use of tall stacks and intermittent controls provided an unreliable means of achieving clean air, the EPA announced in early September 1973 that it would approve their use for some electric power plants. The change in attitude took place, said the EPA, because of the energy crisis, because sufficient scrubbers were not available, and because of the newly improved reliability of intermittent controls.[23] Despite the policy shift, EPA officials still expressed misgivings about the use of such controls. The agency cited the possibility that their use could acidify soil and water, reduce visibility, and increase the formation of sulfates. Paradoxically the EPA's willingness to accept additional sulfates came only a few days before John Finklea and

EPA Administrator Russell Train told congressional subcommittees of their apprehension about these same sulfates.

President Nixon quickly followed the EPA decision with a request for a temporary relaxation of SO_2 emission standards in order to avoid severe fuel shortages during the coming months. Most of the relaxations would be granted in the Northeast and upper Midwest, where the greatest fuel shortages were expected. The variances were not meant to change the air quality standards, only their date of accomplishment. The distinction was not unimportant. To delay implementation of the standards could retard further progress and put a stamp of approval on existing levels of pollution. These points were not lost on Senator Muskie, who criticized President Nixon in September 1973 for abandoning environmental safeguards and for ignoring such alternative solutions as fuel conservation and mandatory allocation programs.[24]

These criticisms had little effect on either President Nixon or on Senator Muskie's congressional colleagues. Some action to stem the energy shortages was necessary, and the choice of solutions had to be made quickly. Several energy bills were introduced in both chambers, and most of them called for a broad-scale relaxation or downgrading of environmental priorities. Among the solutions considered were provisions that would allow temporary suspensions of any fuel or emission limitations for stationary sources, prohibitions on the use of petroleum and natural gas by major fuel users, and mandatory conversion from these fuels to coal. Although these conversions would take place after consideration of health and environmental concerns, primary air quality standards for SO_2 could be violated.[25] Despite his earlier opposition to such a strategy, the situation had deteriorated so much by November 1973 that even Senator Muskie introduced a bill that would have allowed the EPA administrator to authorize temporary suspensions of any emission limitation or timetable. These delays, which would not be subject to public hearings or judicial review, could be granted in all cases except where the administrator determined that "such suspensions will present an imminent and substantial endangerment to the health of persons."[26]

President Nixon left no doubt about his feelings when he stated his intent to "propose some things that will drive the environmentalists up the wall."[27] One of these, which did not require any formal action, was to ignore the EPA as a consultative source on how environmental regulations could be modified in order to cope with the energy problem. Neither President Nixon nor the OMB seemed interested in maintaining environmental controls if they created obstacles to solving the shortage of energy supplies. Russell Train continually stressed that environmental programs were not responsible for the nation's problems, but he found few receptive listeners in other parts of the executive branch.

Train also met with little success among many congressmen who were convinced that clean air and adequate supplies of energy were contradictory goals. The Energy Emergency Act of 1974 reflected this attitude. It required certain power plants to convert to coal, and these plants would have until January 1, 1979, to comply with emission limitations included in state implementation plans.

Another part of the energy bill allowed violations of the primary air quality standards as long as no one would be exposed to "significant health risks."[28] The Congress provided no definition of this phrase other than to say that such risks are less severe than those associated with imminent and substantial endangerment. At best, Congress created confusion by establishing a new and vaguely defined meaning for acceptable levels of air quality.[29] At worst, it was telling the EPA that virtually any amount of additional pollution would be acceptable since no one knew the level of air pollution at which significant health risks occurred. This congressional ambivalence was even more peculiar in light of the purpose of the air quality standards, the conclusions of the Rall report, and the preliminary findings of the CHESS studies. The standards were meant to be protective of the public's health, and the two studies had indicated that the standards for SO_2 were appropriate or perhaps even too lax. Congress decided to sacrifice some degree of public health and previous progress in attaining clean air for some unknown amount of additional energy supplies.

In order to avoid the claim that this choice demonstrated an inadequate sensitivity to the public's health, the Congress authorized the expenditure of $3.5 million for a study on the health effects of sulfur oxides. Like the earlier Rall study, the NIEHS would conduct the new work in cooperation with the EPA. A study to determine the effects of exposure to SO_2 would seem unnecessary in view of existing studies, but the Senate's report on the Energy Emergency Act said that further study would be useful because the "health effects of sulfur oxide emissions from the burning of coal are not now fully known or understood."[30] What the report failed to mention was that these effects would probably never be fully comprehended.

The Energy Emergency Act provided for some relaxation of the environmental rules that Nixon had railed against, but the bill had nonenvironmental provisions that he could not tolerate. As a result, Nixon vetoed the bill on March 6, 1974, and the Senate failed to override the veto later that day. For environmentalists, the veto appeared to be a blessing from an unlikely source, although it would not appear that way in the weeks to come.

Even before Nixon vetoed the energy bill, the OMB was in the midst of developing recommended changes in the Clean Air Amendments in order to alleviate some parts of the energy problem. The proposals that would have affected levels of SO_2 included the following:

1. All power plants would be allowed to use tall stacks and intermittent controls as a permanent means of compliance with air quality regulations regardless of the plants' location, their ability to install scrubbers, or their eventual contribution to higher levels of SO_2 and sulfates.

2. Power plants converting to coal would not have to comply with emission limitations until 1984 (instead of 1979 as first proposed in the Energy Emergency Act).
3. The EPA would be given more discretion in enforcing legal compliance dates for meeting air quality standards.
4. The EPA would be required to consider and balance social, economic, and environmental elements before it set any additional standards.
5. Significant deterioration of pristine areas would be allowed because states would be prohibited from setting standards more stringent than already-existing primary and secondary standards.

Not only were these changes much broader than Nixon had sought a few months before, but in some people's opinion, the changes would have emasculated the Clean Air Amendments' most important provisions. Russell Train was adamantly opposed to many of the changes. He had little chance, however, to convince the president that the changes went too far, especially since they had already been approved by the OMB, the FPC, the CEQ, and the recently created Federal Energy Office. Moreover in the six months that Train had been in charge of the EPA, he reportedly had never met privately with Nixon.[31]

Despite his short tenure, many people thought that Train's resignation was imminent. The administration seemed to view him and his agency as unwanted black sheep. Train's opinions on potential changes in environmental programs were often neglected or not even solicited. From his perspective, he did not want to be seen as the one responsible for the end of much of the improvement that had occurred during the years since the EPA's creation. Nonetheless Train was even in a more awkward situation than Ruckelshaus had been when he considered resigning in December 1972. Unlike Ruckelshaus, Train had never received the president's public support or any public indication that the White House backed the EPA's programs. According to Train, when he met with the president to talk about his appointment as administrator of the EPA, there was no substantive discussion concerning what the agency should do.[32] It seemed that Train, who had asked for the position, was willing to accept it without receiving the same assurances that Nixon had given to Ruckelshaus a few years before. Regardless of these considerations, the resignation did not come, and Train eventually convinced the OMB to delete the proposal on the social and economic variables and to shift the compliance deadline for converting plants from 1984 to 1980.

Some changes were made, but Train was not entirely satisfied. In a highly unusual procedure, Train forwarded the administration's proposed amendments to the Congress but then indicated that he disagreed with the provisions that allowed significant deterioration and the permanent use of tall stacks and intermittent controls. He based his opposition to the latter provision on the fact that the techniques contributed to increases in sulfate levels. He pointed out that adverse health effects occur when ambient levels of sulfates reach 8 to 10 micrograms per cubic meter, and he emphasized that such levels are exceeded regularly in many parts of the Midwest and Northeast.[33]

Train failed to sway the White House on the question of intermittent controls, but he did succeed in convincing the OMB that primary air quality standards should not be violated because of any energy-related conversions to coal. As a consequence of its support for intermittent controls and its new opposition to the violation of primary standards, the administration placed itself in a contradictory position. It was willing to protect the public's health from the adverse effects of SO_2 but apparently not from the more damaging effects of sulfates.

The administration's efforts to achieve a wide-ranging relaxation of environmental regulations ran into difficulty when its proposals came under sharp criticism from many environmental groups and several of the states' environmental officials. In addition, by the time the new proposals were sent to Congress, the coldest months had passed. The nation had survived the winter without a significant relaxation of environmental laws in spite of claims that such changes had been imperative. The crisis had abated, at least temporarily, with the warmer weather and the end of the oil embargo. Finally there was also a strong feeling among many Americans that they had been the victims of a cruel hoax perpetrated by some multinational oil companies. Supposedly the companies had fabricated the oil shortage in an effort to increase their profits.

All of these factors contributed to a bill that provided for far fewer changes in environmental regulations than President Nixon had originally sought or had gotten in the Energy Emergency Act. The Energy Supply and Environmental Coordination Act of 1974, Public Law 93-319, which amended parts of the air legislation, allowed power plants to convert to coal but only under a rigid set of guidelines. Other major provisions affecting sulfur oxides included the following:

1. As a result of an amendment that Senator Muskie introduced, conversions would not be allowed if EPA's administrator judged that violations of primary standards were likely. If the administrator approved a conversion and violations of primary standards subsequently occurred, the EPA could initiate enforcement actions under the provisions of the Clean Air Amendments.

2. In air control regions where standards for a pollutant were already exceeded, conversions would be allowed only if the power plant met existing emission limitations for that pollutant. This provision was also in response to Senator Muskie's efforts.

3. Conversions would be prohibited if they resulted in increases in levels of noncriteria pollutants to the point that these pollutants caused or contributed to a significant risk to public health. The noncriteria pollutants, for which air quality standards did not exist, included sulfates, sulfuric acid,

and trace metals. The Congress inserted this provision in response to reports such as those of Rall and the CHESS researchers that these pollutants are linked to adverse health effects.

4. If the first three conditions were met, emission limitations could be suspended through the end of 1978 for converted plants as long as the affected companies agreed to make a long-term commitment to use low-sulfur coal or to install permanent control devices (scrubbers) by the end of the suspension period. Intermittent controls could be used only during the suspension period if they assured the attainment and maintenance of air quality standards. At Senator Muskie's insistence, the EPA would determine the acceptability of the intermittent controls. By including this provision, Muskie ensured that relatively few intermittent systems would be used in light of the EPA's known opposition to their widespread use. Equally important, for converted plants unable to obtain a long-term supply of low-sulfur coal, permanent controls would be mandatory. This congressional decision was a victory for the EPA and a rebuff to the rest of the Nixon administration, which had favored extended use of tall stacks and intermittent controls. Finally the law again authorized the expenditure of $3.5 million to study the effects of increased SO_2 emissions resulting from conversions to coal.

From an environmental perspective, the new law was far more stringent than the Energy Emergency Act had been. Unlike the earlier bill, the Energy Supply and Environmental Coordination Act maintained the importance of environmental values; it repeatedly stated that the public's health should not suffer because of any energy shortages, and it acknowledged that clean air and energy development were compatible.

The best indication of the differences in the impacts of the two acts are shown in table 7-1. The figures in the table compare three possible scenarios for coal conversion. The first set of figures shows the maximum number of conversions that would have been expected under the provisions of the Energy Emergency Act, which allowed violations of health-related standards and which ignored noncriteria pollutants such as sulfates. With the first scenario, large amounts of oil and natural gas would be saved, but the result would be a huge increase in the amount of SO_2, particulates and, probably, sulfates. Most of the increases in pollution levels would have occurred in Pennsylvania, New York, and New Jersey, three states that could least afford the increases. For these states, the conversions would double the amount of SO_2 emitted from existing power plants.

The second and third situations show the number of conversions expected if primary standards could not be violated or if the term *significant risk* was interpreted to prevent adverse effects associated with sulfates. In fact, the EPA did decide to apply the significant-risk concept to the sulfate problem.[34] Accord-

Table 7-1
Estimated Energy and Environmental Impacts of Coal Conversions, 1977

Scenario	Number of Power Plants Converted	Increased Coal Demand (10³ tons/yr)	Oil Savings[a] (10³ bbls/yr)	Gas Savings[a] (10⁶ cf/yr)	Additional SO₂ Emissions (tons/yr)
1. Maximum conversions, no environmental constraints	72	21,869	65,079	79,360	472,591
2. Violation of primary standards prohibited	19	6,030	18,805	15,804	208,875
3. Significant risk from sulfates prohibited	5	983	2,456	6,514	35,150

Source: Federal Energy Administration, *Final Revised Environmental Impact Statement: Coal Conversion Program*, FES-77-3 (Washington, D.C.: Government Printing Office, 1977), chap. 4.
[a]For oil and natural gas, the units are barrels and cubic feet, respectively.

ing to the environmental agency, emissions of SO_2 would be restricted in areas already reporting high concentrations of SO_2, sulfates, or particulates.

Notes

1. Anthony Downs, "Up and Down with Ecology—The 'Issue-Attention Cycle,' " *Public Interest* 28 (Summer 1972): 38.

2. *Federal Register* 36 (April 7, 1971): 6692

3. EPA, *National Public Hearings on Power Plant Compliance with Sulfur Oxide Air Pollution Regulations* (1974), p. 4.

4. CEQ, *Environmental Quality—1973* (1973), p. 162.

5. U.S. Senate, Committee on Public Works, Subcommittee on Air and Water Pollution, hearings, *Implementation of the Clean Air Amendments of 1970, Part 2*, 92d Cong., 2d sess., 1972, pp. 786-787.

6. U.S. Senate, Committee on Public Works, Committee Print, *Environmental Protection Affairs of the Ninety-third Congress*, 94th Cong., 1st sess., 1975, p. 12.

7. These percentages were computed from data provided to the author by the FPC.

8. U.S. Senate, Committee on Public Works, Subcommittee on Air and Water Pollution, hearings, *Nondegradation Policy of the Clean Air Act*, 93d Cong., 1st sess., 1973, p. 81.

9. Louis Harris and Associates, *The Harris Survey Yearbook of Public Opinion, 1970* (New York: Louis Harris, 1971), p. 46, and "Economy Tops List of Concerns Again," in "Harris Survey," January 1, 1976.

10. Gladwin Hill et al., *Protecting the Environment: Progress, Prospects, and the Public View* (Washington, D.C.: Potomac Associates, 1977), pp. 22, 24.

11. Cited by Congressional Quarterly, *Continuing Energy Crisis in America* (Washington, D.C., 1975), p. 15.

12. See *New York Times*, March 12, 1974, p. 12. The data on spending preferences are from the National Opinion Research Center and are reported in Department of Commerce, *Social Indicators: 1976* (1977), p. lxxvii.

13. John Quarles, *Cleaning Up America* (Boston: Houghton Mifflin, 1976), pp. 207-208.

14. U.S. Senate, Committee on Interior and Insular Affairs, hearings, *Energy Emergency Legislation*, 93d Cong., 1st sess., 1973, pp. 271-279.

15. The President's "Message to the Congress on a Program to Insure an Adequate Supply of Clean Energy in the Future," is reprinted in U.S. House of Representatives, House Document 118, 92d Cong., 1st sess., 1971.

16. *Public Papers of the Presidents of the United States: Richard M. Nixon—1973* (1974), p. 102.

17. Ash's letter is reprinted in the *Congressional Record* 119 (September 6, 1973): 28684–28685.

18. *Environmental Health Letter* 12 (August 15, 1973): 2.

19. PL 90-148, sec. 305(a).

20. Rall's report is reprinted in National Research Council, *Air Quality and Automobile Emission Control*, vol. II: *Health Effects of Air Pollutants*, a Report by the Coordinating Committee on Air Quality Studies, NAS, National Academy of Engineering for the U.S. Senate, Committee on Public Works, Committee Serial No. 93-24, 93d Cong., 2d sess., 1974, pp. 413–473.

21. Ibid., p. 421.

22. *Congressional Record* 119 (September 6, 1973): 28682.

23. *Federal Register* 38 (September 14, 1973): 25697–25700.

24. *Congressional Record* 119 (September 10, 1973): 29030.

25. See, for example, U.S. Senate, Committee on Interior and Insular Affairs, Report No. 93-498, *National Energy Emergency Act of 1973*, 93d Cong., 1st sess., 1973, pp. 22–23.

26. S.2680, 93d Cong., 1st sess., 1973.

27. *Washington Post*, December 22, 1973, p. A1.

28. U.S. Senate, Report No. 93-681, *Energy Emergency Act*, 93d Cong., 2d sess., 1974, p. 90.

29. A few months after the Congress included the significant-risk concept in the bill, Russell Train admitted that he did not know what it meant. "It is a new concept," he said. "It is neither a primary standard nor secondary standard and does inject an ambiguity and an uncertainty." See U.S. House of Representatives, Committee on Interstate and Foreign Commerce, hearings, *Standby Energy Emergency Authorities Act*, 93d Cong., 2d sess., 1974, p. 101.

30. Senate, *National Energy Emergency Act of 1973*, p. 30.

31. *Wall Street Journal*, March 22, 1974, p. 1.

32. U.S. Senate, Committee on Government Operations, Committee Print, *Study on Federal Regulation*, vol. I: *The Regulatory Appointments Process*, 95th Cong., 1st sess., 1977, p. 21.

33. Train's comments are contained in a letter, dated March 22, 1974, to Honorable Gerald R. Ford, President of the Senate. The letter is reprinted in *Congressional Record* 120 (April 2, 1974): 9190–9192.

34. EPA, Office of Air and Waste Management, *Position Paper on Regulation of Atmospheric Sulfates* (Research Triangle Park, N.C., 1975), p. xix.

8

Continuing Debate

The energy situation in late 1973 and early 1974 had provided an unparalleled opportunity for assault on the air quality legislation. Although the attacks had not succeeded as much as some people had hoped, this did not cause them to cease their opposition. As the discussion in previous chapters has demonstrated, the merits of regulatory policies are the subject of constant discussion because of changed social conditions, the election of new officeholders, the appointment of new administrators, and so forth. As a result, modification of regulations is always possible.

For the opponents of the clean air legislation, there were several added reasons to persevere. First, even a slight weakening of environmental regulations could save industry millions of dollars. Second, the 1970 legislation was not intended to be permanent, which meant that new legislation would have to be passed, and more hearings, negotiations, and compromises would be required. Oversight hearings would also be conducted. Each of these steps would provide interest groups with many more chances to assert their positions and to affect policy making. Third, as a result of the energy situation, new studies on sulfur oxides had been initiated, and each of these created the possibility of revised criteria or standards and further pressure on the EPA. In short, all these reasons suggested that continued opposition would be worthwhile and perhaps even beneficial to affected economic interests.

External Support for the EPA

Throughout the early 1970s, the EPA faced a familiar pattern of opposition from other federal agencies concerned with protecting the interests of their constituents. This was not the case in the summer of 1974, however, when two groups released reports relevant to the SO_2 standards. The Federal Power Commission (FPC) produced the first of these. The commission, charged with ensuring an adequate supply of electricity, had initiated a study to consider alternatives to the use of oil. As part of the work, one task force had examined the likely health impacts related to the use of such high-sulfur fuels as coal. Relying heavily on the CHESS data and John Finklea's assistance, the task force predicted that if such fuels were used and primary standards were violated as a result, the adverse health effects could be immense. These would include thousands of premature deaths and great discomfort to the youngest and oldest members of the population.[1]

In addition to providing support to those opposed to widespread conversions to coal, the report also lent credence to the EPA's latest data on the relation between sulfur oxides and ill health. Only rarely had the EPA and the FPC agreed on major issues affecting energy and the environment and, in many instances, the commission had been one of the EPA's chief antagonists. Hence the conclusions of the task force represented an unexpected but successful accomplishment for the EPA.

Shortly after the FPC published its report, the National Academy of Sciences (NAS) completed its own four-volume study, which had been initiated at the request of the Senate Committee on Public Works. The NAS is a private, official, but independent adviser to the federal government on matters of science and technology.

Although the original focus had been on automobile-related pollutants, the NAS soon expanded its efforts to include sulfur oxides and particulates. When the study was finished in August 1974, its discussion of these two pollutants offered little that was novel. While expressing dissatisfaction with the adequacy of the data base for all air quality standards, the academy concluded that existing standards for the six criteria pollutants need not be changed on the basis of recently acquired evidence. In this respect, the report took issue with the CHESS studies' conclusion that the short-term SO_2 standard might not be sufficiently stringent. As the NAS put it, there was considerable doubt about the validity of such a conclusion.

The difference of opinion may have been merely a question of semantics because the NAS went on to indicate that the margins of safety built into the standards were both small and potentially misleading. The margins were small compared with those used for other environmental pollutants, such as radiation. The use of margins could be misleading, however, because they are based on the assumption that scientists can establish a threshold level below which no harm occurs. As the NAS explained it, each person has a different threshold level depending on such factors as age, health, personal life-style, and so forth. For some people with heart and lung disorders, the difference between safe and harmful levels might be so small "that the slightest increase in pollution could aggravate their condition or precipitate death."[2] In sum, the existing margins of safety were said not to "guarantee that such [adverse] health effects are completely absent even at the level of the ambient air quality standard."[3]

One further disquieting finding concerned the occurrence and impact of synergistic reactions among pollutants. The report noted that these reactions generally need not be a source of concern except in cases where SO_2 and particulates interacted to produce more toxic sulfuric acid or particulate sulfates. Although an understanding of the relation between SO_2 and particulates would benefit from additional research, the NAS stressed a very important point: air quality standards set on the basis of individual pollutants "can never fully reflect ambient pollutant concentrations and their effects on human health."[4]

Immediately after the NAS finished its study, Senator Randolph asked that it prepare a more comprehensive analysis of sulfur oxides and techniques for controlling them. In many respects, the second report, which was finished in March 1975, was similar to the first. Unlike the earlier report, however, the newer study concurred with the EPA's judgment that measured levels of sulfates between 8 and 12 micrograms per cubic meter could lead to adverse effects among elderly people with heart and lung disorders. The NAS reviewers also agreed that sulfates are probably more toxic than SO_2 alone. With this fact in mind, the second report concluded that it is "inappropriate to use the failure to observe a human response to sulfur dioxide during controlled [laboratory] exposure as a reason to allow an increase in the emission of sulfur dioxide."[5]

In another indication of support for the EPA, the NAS endorsed the widespread use of scrubbers for power plants within three hundred miles upwind of major cities. Tall stacks and intermittent controls, in contrast, were not recommended except in carefully defined, short-term situations. The NAS expressed apprehension about the use of both strategies because they did not decrease the total amounts of either SO_2 or its sulfate by-products.

From an energy perspective, both the FPC and the NAS studies raised serious doubts about efforts to increase substantially the amount of high-sulfur coal burned in the United States. There was no quibbling with any of the three reports about the need to meet the primary air quality standards. Sufficient agreement existed even on the question of sulfates to justify a particularly cautious approach to energy-related conversions. The likelihood of just such an approach was enhanced in view of the NAS's finding that the benefits of reducing emissions of sulfur oxides probably exceeded the costs substantially for power plants located in areas already reporting high levels of SO_2. In addition, the report placed a special emphasis on prudence when it noted that "the consequences of an error in judgment which led to substantial damage to human health would be more serious than an error which led to economic misallocation."[6]

The EPA's obligation under the Energy Supply and Environmental Coordination Act to limit coal conversions if significant risks to health would result from noncriteria pollutants also seemed to ensure the application of a prudent approach. In addition to its own CHESS studies, the EPA could now point to data from outside the agency to support its health-based efforts to halt or minimize the number of environmentally unsound coal conversions.

President Ford and the FEA

The change in presidents in August 1974 did not mean a change in attitudes about environment protection. President Gerald Ford seemed at least as committed to a relaxation of environmental standards as his ill-fated colleague had been, and many people felt that Ford was actually more opposed to environmental controls than Nixon had been.

Once in office, Ford wasted little time in proposing solutions to the nation's energy problems. He wanted to achieve a significant increase in the number of conversions to coal and even a larger number than Nixon had called for. As an example, Ford initially expressed a desire to switch all oil-fired power plants to the use of coal by 1980. After he had been in office for a few months, the president presented the Congress with specific proposals for changes in the Clean Air Amendments, and he followed this action with a program for energy independence a few weeks later. Among the changes that Ford sought were proposals intended to eliminate any restrictions on significant deterioration, to extend the deadline for mandatory installation of scrubbers from 1979 to 1985, and to permit certain other plants to use intermittent controls through 1985.

Although the problems of increased SO_2 emissions could not be ignored, it was clear that some solution to the sulfate issue would have to be found. In the absence of an adequate understanding of how sulfates are formed, it would be impossible to develop an appropriate control strategy, at least in the short run. Efforts to implement large-scale conversions to coal would be politically risky and would surely be subject to harsh criticism from environmental groups concerned about the possibility of significant risks to the public's health. Thus the emerging data on the severity of the sulfate issue loomed as a major stumbling block. Alternatively if it could be shown that the studies on sulfates were inaccurate or that they had reached improper conclusions, major progress could be made, or so it was believed.

At this point, the Federal Energy Administration (FEA) entered the scene. This agency, which had been created in May 1974 as a successor to the Federal Energy Office, was supposed to implement the nation's energy policies and generally strive to ensure the availability of adequate fuel supplies. Although the FEA was not created to oppose environmental values, its primary missions would certainly conflict with many environmental programs.

One of the FEA's first responses to the many sulfate studies was its establishment of the Sulfate Scientific Review Panel in October 1974. One of the panel's major purposes would be to evaluate and review the studies so that the president's energy program could proceed. The panel was intended to be a cooperative venture between the FEA and the EPA, but the EPA opposed such a review in the context of the coal conversion program. The environmental agency was willing, however, to establish a panel of outside experts to conduct an independent assessment of the sulfate question. The FEA initially agreed to this arrangement, and the EPA convened a group of experts acceptable to the FEA.

Despite the apparent agreement, the FEA decided to hire its own consultant to provide it with information about the sulfate problem. As a reading of FEA documents and correspondence shows, the agency left little doubt about what it hoped to receive from the consultant, who was asked to discuss the amount of uncertainty and the reliability of EPA's data base, outline the "scientific uncertainties and technical weakness involved in making national estimates from dose-response functions derived from epidemiological evidence," and explain any obvious weaknesses in the EPA's methods.[7] While this is not a comprehensive listing of what the FEA wanted, the wording of these three tasks does reflect the energy agency's predilections.

When the contractor finished the report, "A Critical Evaluation of Current Research Regarding Health Criteria for Sulfur Oxides," the FEA presented the study to the relevant environmental committees in the Congress. Most of the work had been done in a very short period, but this did not detract from its appeal to the FEA. The consultant provided the FEA with exactly what it seemed to want: a short, highly readable study that directly challenged the EPA's scientific conclusions and supported the Ford administration's efforts to weaken the Clean Air Amendments. The report severely criticized much of the EPA's research findings on sulfur oxides and suggested that many of them were without scientific foundation. A sampling of the study's conclusions illustrates their fundamental disagreement with those of the EPA:

1. There is no valid evidence showing that SO_2 or sulfates are a cause of mortality.
2. The available data do "not support a cause-effect relationship between adverse health effects and sulfur oxides."
3. The extent to which air pollution must be reduced to avoid ill effects is still not certain.
4. The EPA's scientific data do "not support conclusively [its] current position . . . with respect to the health effects of sulfur oxide emissions."
5. It is inappropriate to say that sulfates are a major cause of ill effects. Thus the promulgation of an air quality standard for sulfates would be "unwarranted and unsubstantiated in the light of our present knowledge."[8]

Criticism of these conclusions did not take long to surface. Senator Muskie called attention to the possibility that the FEA report might undermine the public's confidence in the air quality standards and, in turn, encourage polluters to seek delays and postponements from compliance dates. He was also concerned that the FEA would use its study instead of those of the EPA to explain the environmental impacts of coal conversions. Senator Randolph was much less charitable to the FEA. He wanted to know why the FEA had ignored the studies on sulfur oxides that the NAS had provided.

FEA Administrator Frank Zarb replied that work on the FEA study had begun before he had assumed control of the agency, and he did not want to assume responsibility for the project. Nonetheless Zarb claimed that it contained information of great importance. He had not intended that it be used unfairly; in his own words, introduction of the report was not intended "to be a low inside curve."[9] Regardless of the FEA's purposes, the reaction in Muskie's subcommittee ensured that the report would have little impact on any forthcoming legislative action.

The subcommittee's unfriendly response did not deter the FEA from further criticism of the EPA or its concern for sulfates. During the summer of 1975, the FEA published a short pamphlet whose major purpose was to encourage the use of intermittent control systems and to rebut EPA's charges about sulfates.[10] On the first count, the energy agency said that intermittent systems are a reliable means of meeting air quality standards, as well as being much less expensive than scrubbers. On the second count, the FEA questioned the EPA's apprehension about and scientific data on sulfates. To the FEA, available studies had not substantiated the claim that sulfates are more harmful than SO_2. Consequently the FEA concluded that it would be inappropriate to reduce emissions of SO_2 below the level necessary to achieve the existing air quality standards. Finally the FEA acknowledged that it was unopposed to the regulation of sulfates as long as such regulation did not precede the availability of all facts. Apparently the best scientific judgments of the EPA and NAS would not be sufficient. The FEA wanted incontrovertible data before it would consent to the regulation of sulfates or a prohibition on intermittent controls.

Unfavorable Responses from Industry

The FEA's reports, which apparently did not sway congressional opinion, were only a small part of a larger effort by trade and industrial concerns to alter the Clean Air Amendments. During the worst of the energy crisis, much of the discussion about how to solve the nation's energy shortages had focused on means to relax environmental standards. This did not mean, however, that industry had accepted the validity of the standards, which the EPA had established without much apparent consultation with affected groups. Thus it was not surprising that the American Iron and Steel Institute complained in the spring of 1974 at one congressional hearing that air quality standards for many pollutants are based on inadequate or inappropriate data. The solution to this problem, concluded the steel lobby, would be to take the standard-setting function away from the EPA and give it to the National Academy of Sciences. According to the plan, the NAS would establish an advisory panel of affected groups so that "consensus standards" would be developed.[11] Unlike the existing standards, consensus standards apparently would be subject to polluters' approval and willingness to meet them. A few months after it made this suggestion, the steel group recommended that an environmental review board be authorized to review and approve proposed air quality standards. The EPA would not be represented on the board, but the OMB, the Federal Energy Administration, and the Departments of Commerce and Interior would be. Their chore would be to review standards to determine the need for them, their possible attainability, and their justifiability in terms of their social, economic, and environmental impacts.[12]

The suggestion that a review board be created surely reflected disenchantment with the standards themselves. It also demonstrated the feelings of many regulated interests, however, that the EPA was too insulated from and uncooperative with them. This represented a pattern of relations with which many interests were unfamiliar and uncomfortable. They did not like being excluded from the processes of policy implementation.

During Senate and House hearings conducted in 1975, other complaints from industry focused on the vague legislative direction provided in the Clean Air Amendments. One company wanted the Congress to define what was meant by the "public health" and another called "an adequate margin of safety" a meaningless phrase. Shell Oil compared the law's margin-of-safety requirement to a surgeon who examined a patient's infected finger and then decided to amputate at the shoulder instead of the elbow, just to be safe. Shell also requested that the Congress provide "appropriate statutory definitions of each word, term or phrase contained in the Act's statement of purpose."[13] Interestingly these complaints were directed at the deliberate vagueness that ordinarily guarantees the access of interest groups to the arenas of administrative decision making. Obviously such was not the case with the EPA, and the industries felt compelled to ask the Congress for a clearer explanation of its intent because of the industries' strong distaste for the EPA's interpretation.

Because the EPA had begun to focus on the health effects of sulfates, the electric utilities also addressed this issue during congressional hearings. Their attack was strongly reminiscent of their stance on SO_2 in the 1960s. In brief, the industry's representatives contended that because insufficient evidence existed to incriminate sulfates, the best policy would be to avoid any control programs until compelling data could be collected and analyzed. This work could take at least five to ten years, and even that period might not be adequate to resolve the debate, or so it was argued. As one lobby for the utility industry suggested, "Even if subsequent research proves that one or more sulfates are major contributors to adverse health effects, the matter will not be settled."[14]

This argument was used to buttress the case for intermittent controls. According to the industry, these controls would provide a more effective and cheaper means of ensuring compliance with air quality standards than would such permanent controls as scrubbers. If intermittent controls actually increased the levels of sulfates, this result might not be so bad, implied a spokesman for the utilities, because it could provide more data for analysis. In any case, the industry asserted that it should be able to select any control strategy as long as it ensured attainment of the standards. If further research found sulfates to be harmful, then the industry would comply with whatever regulations were necessary to prevent harm. Until that time, however, the industry argued that the EPA should not force utilities to install scrubbers, especially since they would cause consumers to pay billions of dollars for environmental equipment that might not provide any corresponding benefits, might not prevent harm from sulfates, and might actually be detrimental.[15] In sum, the argument seemed to be that regulation would be premature in the absence of indisputable data. Once again, the arguments against environmental regulations had barely changed from the early 1960s to the mid-1970s.

The EPA's Answer

It is uncertain exactly how the criticisms affected the EPA, but the agency did announce in mid-1974 that it planned to assess periodically the validity of all the standards, with the first review to be completed in the following two years. The review would be a routine scientific exercise and was not "prompted by any lack of confidence in the standards."[16] For many people, EPA's statement was an attempt to delay what it should have already done. The 1970 legislation required the agency's administrator to conduct periodic reviews of all criteria and to modify or revise them as necessary. The document for sulfur oxides, which had been published in early 1969, was based on data collected mainly in the 1950s and early 1960s and nothing later than mid-1968. In mid-1974, then, most of these data were already more than six years old, and the oil, coal, steel, and electric utility industries believed that more recent findings on SO_2 justified revision of the criteria and subsequent relaxation of the standards.

In answer to complaints about energy shortages, Administrator Train largely discounted claims that implementation of the Clean Air Amendments was responsible for the country's energy problems. Under no circumstances, Train emphasized, should changes in other national priorities, such as energy, be used to dismantle environmental safeguards. Indeed he observed that such changes made it imperative to "insist even more strongly upon rigorous standards and safeguards."[17]

One place where further safeguards seemed appropriate was in the area of sulfates. Train repeatedly told members of the environmental subcommittees in both legislative chambers that sulfates represented a serious health risk and a far greater problem than SO_2 alone. He concluded that the standards for SO_2 should be supplemented or replaced with a standard for sulfates. Scientific understanding of sulfates was still incomplete, but, as Train told members of the Senate Subcommittee on Environmental Pollution, "The public health carries with it too high a priority to permit us to simply sit back and wait until we do have all of the answers." He recognized the "need to proceed with some type of control action in the absence of fully definitive data."[18] Train cited the sulfate problem as one example of this kind of situation.

Concern for sulfates was heightened in the summer of 1975 when a House subcommittee conducted four days of hearings on the issue and when the EPA published a special report on atmospheric sulfates. In the House hearings, not much new information was developed, but several witnesses stressed their discomfort with the adequacy of the existing SO_2 standards. Much of the dissatisfaction was based on the feeling that they did not offer protection from sulfates. A summary of the testimony concluded that a real need existed for an air quality standard for sulfates and sulfuric acid.[19]

In its report, the EPA said that its best judgment was that adverse health effects could be expected when levels of sulfates are as low as 6 to 10 micrograms per cubic meter over a period of twenty-four hours and 10 to 15 micrograms per cubic meter over a year.[20] This finding was especially alarming in view of the fact that in the early 1970s, monitoring stations had recorded annual levels of sulfates ranging from 10 to 24 micrograms per cubic meter in the urban areas of nearly half the states, including Ohio, Indiana, Illinois, Massachusetts, Pennsylvania, and New York. Prudence was the key, because the EPA recommended a strategy that would limit increases in SO_2 in order to minimize the formation of sulfates.

Train's testimony before the Congress, the tenor of the House hearings on sulfates, and the EPA report all gave the impression that regulatory action on sulfates was both essential and imminent, but this was not the case. The EPA said that it had "a great deal of scientific support" for its conclusions about sulfates but not enough to set an air quality standard. Train estimated that at least five more years of study would be necessary before acceptable levels of sulfates could be determined.

At the least, the EPA's position on sulfates put the agency in a difficult situation. Train had argued that regulatory action should not be held in abeyance pending the availability of complete information, but he then proceeded to ignore his recommendation. He said that a decision on what to do about sulfates was several years away, but his position ignored highly salient considerations. A decision to delay regulatory action in the face of inconclusive data "is in reality a decision to permit continued exposure to what is now thought by some to be a higher level of risk."[21] Thus the EPA's failure to initiate some form of regulatory control sanctioned further increases in sulfates, with the concurrent increase in health risks. His stance also reflected the EPA's assessment that research to be completed in the future justified delay in the present.[22] Such a judgment assumes that as-yet-uncompleted research will be both productive and definitive, but as all scientists know, this assumption can easily be faulty.

The EPA's reluctance to set a standard for sulfates and its continuing disinclination to do so in the late 1970s is perplexing in light of other events that occurred in the early and mid-1970s. When the EPA first established air quality standards in 1971, it had taken the position that if gaps existed in supporting data, subsequent research would be conducted to fill them. In the meantime, however, the EPA had insisted that incomplete scientific knowledge did not justify delay.[23] This position was entirely consistent with that of the Senate, which had stated in its report on the 1970 amendments that the best information available could be used to set standards.[24] The EPA's policy in 1975 was just the opposite. An incomplete understanding of sulfates and their effects was said to justify a delay in the standard-setting process, even though the EPA readily conceded that sulfates "are more likely than sulfur dioxide alone to be responsible for many of the adverse health effects typically associated with sulfur oxide."[25]

In addition to the administrative precedent, at least two court decisions seemed to provide justification for action on the sulfate issue. The first case involved EPA regulations designed to reduce the lead content of gasoline. Although the agency did not have complete scientific data, it had concluded that leaded emissions pose a significant risk to the public's health. When challenged in the courts, the EPA successfully argued that conclusive proof of harm was not a necessary precondition of regulatory action.[26]

In a second case of even greater relevance, a federal district court ruled in March 1976 that once the EPA had determined that airborne lead has an adverse effect on health and that it comes from diverse sources, the agency had a mandatory duty to propose air quality standards for the pollutant in accordance with the provisions of the Clean Air Amendments. In the case, the court rejected the EPA's contention that incomplete data on lead justified the agency's decision not to proceed with the standard-setting process.[27] After an appeal, which the EPA lost, an air quality standard for lead was eventually issued, with the EPA's acknowledgment "that there are still key aspects of the scientific knowledge about lead which are unknown or controversial."[28]

The situation with sulfates is very similar to that of lead. The EPA conceded in 1975 that sulfates come from diverse sources and have an adverse effect on health.[29] In response to this recognition, the Sierra Club requested in January 1976 that the EPA initiate the steps necessary to set a standard for sulfates. Without denying that sulfates are harmful, Train declared that he did not have to set a standard in the face of incomplete data. After Train's refusal to act, the Sierra Club filed a suit seeking judicial review of the administrator's decision.[30]

At first glance, the EPA's recalcitrance in setting a standard for sulfates is difficult to explain. The agency was established with the express purpose of protecting Americans from environmental pollutants. While the decisions required to accomplish this goal might be controversial, it was expected that the EPA would not be reluctant to make them. Moreover the 1970 legislation put the highest priority on the protection of the public health. Other values were accorded lower priority or ignored altogether. Thus some movement in the direction of a standard for sulfates seemed appropriate.

If one looks at the sulfate problem in greater detail, however, the EPA's position is better understood. Sulfates come in many forms, and each has potentially distinct effects on health and welfare. The large variety of sulfates means that separate monitoring procedures and air quality standards might be required for each sulfate. Even if the EPA could establish appropriate standards, it still might be exceptionally difficult to achieve them. No direct relation exists between emissions of SO_2 and levels of sulfates, and the oxidation of SO_2 into the various sulfates is affected by such considerations as temperature and humidity, the intensity of solar radiation, and the presence of other pollutants. Since the EPA cannot control these variables, the agency could find itself in a situation where it could set air quality standards for sulfates but not be able to indicate how the standards could be achieved. In short, each of these considerations created the specter of a regulatory nightmare for the EPA.

From an organizational perspective, other reasons can be provided to explain the EPA's sluggishness. Limited political support, bureaucratic inertia, and anticipation of adverse reactions from polluters are all likely explanations. In addition, the regulation of sulfates could have an unfortunate and inestimable impact on the EPA. The promulgation of air quality standards for sulfates would greatly increase the pressure to rescind the standards for SO_2, which might no longer be necessary. Such a move would confirm the suspicions of many people that the agency had been regulating the wrong pollutant (SO_2). Such an admission would cause extraordinary political and economic dislocation, and the EPA might never regain its credibility.

Notes

1. FPC, *Power Generation: Conservation, Health and Fuel Supply* (1974), a report to the Task Force on Conservation and Fuel Supply, p. 29.

2. National Research Council, *Air Quality and Automobile Emission Control, Summary Report*, a report by the Coordinating Committee on Air Quality Studies, National Academy of Engineering for the U.S. Senate, Committee on Public Works, Committee Serial No. 93-24, 93d Cong., 2d sess., 1974, 1:59.

3. Ibid., p. 18.

4. Ibid., p. 19.

5. National Research Council, *Air Quality and Stationary Source Emission Control*, a report by the Commission on Natural Resources, NAS for the U.S. Senate, Committee on Public Works, Committee Serial 94-4, 94th Cong., 1st sess., 1975, p. 2.

6. Ibid., p. xxxvi.

7. U.S. Senate, Committee on Public Works, Subcommittee on Environmental Pollution, hearings, *Implementation of the Clean Air Act—1975, Part 1*, 94th Cong., 1st sess., 1975, pp. 351–354.

8. Ibid., pp. 399–402.

9. Ibid., p. 296.

10. FEA, *Electric Utilities, Clean Air Act Amendments, and Sulfates* (July 1975).

11. U.S. Senate, Committee on Public Works, Subcommittee on Environmental Pollution, hearings, *Clean Air Act Oversight, Part 1*, 93d Cong., 2d sess., 1974, pp. 216–217.

12. Ibid., p. 231.

13. U.S. House of Representatives, Committee on Interstate and Foreign Commerce, Subcommittee on Health and the Environment, hearings, *Clean Air Act Amendments–1975, Part 2*, 94th Cong., 1st sess., 1975, p. 1245.

14. Senate, *Implementation–1975, Part 2*, p. 1439.

15. Ibid., pp. 1439, 1443.

16. Senate, *Clean Air Act Oversight, Part 2*, p. 1075.

17. Senate, *Implementation–1975, Part 1*, p. 141.

18. Ibid., pp. 155, 166.

19. U.S. House of Representatives, Committee on Science and Technology, Subcommittee on the Environment and the Atmosphere, Committee Print, *Summary of Hearings on Research and Development Related to Sulfates in the Atmosphere*, 94th Cong., 1st sess., 1975, p. 3.

20. EPA, Office of Air and Waste Management, *Position Paper on Regulation of Atmospheric Sulfates* (Research Triangle Park, N.C., 1975), p. x.

21. Rita Ricardo Campbell, *Food Safety Regulation* (Washington, D.C.: American Enterprise Institute, 1974), p. 29.

22. National Research Council, *Decision Making in the Environmental Protection Agency* (Washington, D.C.: NAS, 1977), p. 67.

23. *Federal Register* 36 (April 30, 1971): 8186.

24. U.S. Senate, Committee on Public Works, Report No. 91-1196, *National Air Quality Standards Act of 1970*, 91st Cong., 2d sess., 1970, p. 11.

25. EPA, *Atmospheric Sulfates*, p. vii.

26. *Ethyl Corp.* v. *EPA*, 8 E.R.C. 1785 (1976).

27. *Natural Resources Defense Council* v. *Train*, 8 E.R.C. 1695 (1976).

28. *Journal of the Air Pollution Control Association* 28 (February 1978): 162.

29. EPA, *Atmospheric Sulfates*, p. 64.

30. Before a decision could be reached, the Sierra Club asked the court to dismiss the case in late 1977 because of the changed administration in Washington and because of new statutory guidelines under the Clean Air Amendments of 1977, PL 95-95.

9 Increasing Skepticism

Much of the criticism directed at the EPA in the early 1970s concerned the adequacy of its data. Questions focused on whether they actually justified the standards for SO_2. While admitting that its data were imperfect and that more research would be necessary, the EPA took the position that existing information did justify the health-related standards. Under no circumstances, said the EPA, did the data suggest that the standards should be relaxed.

By 1976, however, much of the EPA's previous success in defending the adequacy of its scientific data dissipated. Within a few months' time, the EPA would find its scientific objectivity impugned before a congressional hearing, its SO_2 criteria in need of a comprehensive revision, and its legislative overseers calling for important changes in its operations.

CHESS: The Final Throes

In the early 1970s, the EPA's confidence in its scientific data was frequently translated into testimony before congressional committees. Perhaps the best example of this tendency was the EPA's use of the CHESS studies, which were referred to in testimony in every year after the EPA's creation. The CHESS findings thus had a major influence on the EPA's activities and the positions it took on such issues as tall stacks and the desirability of coal conversions. At the same time, however, the report's influence was not limited to the environmental agency. The FPC, NAS, CEQ, and various congressional committees all had relied on the CHESS findings in their own reports and analyses.

The report's widespread use did not mean that it was universally applauded. As an example, the CHESS studies has instigated the FEA's critical assessment of the EPA's conclusions on sulfur oxides. Frequent criticisms also came from industrial groups that would feel the burden of increased regulation should the EPA decide to limit further the emissions of SO_2 because of the CHESS studies. The studies also provoked debate within the scientific community. When the EPA first distributed draft copies of the results, it had encouraged widespread review by outside experts. Many of the reviewers identified procedural or methodological problems, but few people questioned the studies' overall conclusions. The EPA itself had formed an expert panel to examine the study, which found weaknesses in the CHESS studies, including problems with the selection of respondents and study areas and difficulties with the accuracy of monitoring devices. Nevertheless the review panel concluded that the studies had made a "very significant contribution to our health information about atmospheric pollutants" and that they had provided evidence for the validity of the SO_2 standards.[1]

With a few exceptions, like the FEA report on sulfur oxides, discussion of the CHESS studies in the early 1970s occurred within the scientific community and remained on an objective and professional level. Rarely did the debate spill over into the popular news media. This situation changed dramatically in 1976. At the end of February, an investigative reporter for the *Los Angeles Times* wrote that John Finklea, who had been in charge of the project, had "systematically distorted" the data in order to prove that sulfur oxides were harmful.[2] According to the reporter, Finklea was guilty of intentional bias, scientific dishonesty, and systematic error. After his subordinates had written individual chapters, Finklea supposedly had rewritten sections to coincide with his own beliefs, which were said to be in conflict with the available data. To accomplish this goal, Finklea was charged with having discarded statistical analyses that he did not like and having included others that were "dubious or unsupportable." The reporter questioned the merits and veracity of the CHESS studies and asked if the EPA should be trusted with the responsibility for further research on air pollution.

These were very serious charges that could not be ignored. If they were accurate, the EPA might never recover its scientific and political credibility. Similarly proponents of environmental protection would find it increasingly difficult to support the EPA if they found that they agency had made some decisions on the basis of tainted data. Several groups and individuals immediately seized on the possibility. Carl Bagge of the National Coal Association raised the possibility of what he said was an environmental Watergate and called for an immediate congressional investigation. Congressman Barry Goldwater, Jr. (R.-Calif.) claimed that his own investigations had served to increase his feeling that there was "a clear factual basis for the allegations."[3] Goldwater later told his congressional colleagues that probable cause existed to "launch an intensive investigation of a possible fraud that has cost the American consumer and taxpayer billions of dollars."[4]

The *Times*'s article and Goldwater's charges came at an inopportune time for the EPA. The House Commerce Committee was then in the midst of considering changes to the clean air legislation, and the allegations could increase the difficulties in arguing against a relaxation of environmental controls. Moreover if the EPA was going to retain its reputation, the charges would have to be rebutted. Two subcommittees in the House decided to provide the EPA with just such an opportunity when they convened a special hearing in April 1976.

Among those who testified, the most supportive of the charges was a scientist who once had worked for the EPA, although not directly with the CHESS researchers. In Robert Buechley's words, the CHESS scientists had promised to policy makers "that specific scientific findings would be forth-coming to incriminate those pollutants for which regulations were being writ-ten."[5] He added that only "positive results were acceptable," and this meant that many qualifying phrases had to be deleted from written summaries in order to prove the researchers' personal biases. Buechley left no doubt about who he thought should receive blame for these scandalous practices: "Dr. Finklea's rather leapfrog rise to a much higher position implies that he was satisfying his superiors in producing a product which he was hired to do."[6]

Other than Buechley, virtually all of the people who spoke repudiated the *Times*'s attack on Finklea. Finklea's former colleagues admitted that the short period they had been given to write the monograph had been a trying period, but they repeatedly supported his integrity and objected to the attacks on him. When the hearing was finished, many minds had been changed about Finklea's honesty. Even Goldwater admitted that his previous concern about Finklea had been allayed. No one who testified detailed a single instance of improper profes-sional behavior on his part, and no one offered any evidence that Finklea had intentionally misrepresented any data.

Although the attacks on Finklea's reputation had been repudiated, the *Times*'s article had also stated that the project itself was faulty. Speaking to this criticism, many people pointed out that the draft report did have problems, but that most of them had been eliminated before publication of the final report. The consensus of those who testified was that the project had produced many valuable data. Russell Train defended the project and tried to clarify how his agency had used its results. Many people felt that the problems with the CHESS data undermined the EPA's previous decisions, but Train replied that in every instance the data had been used as supplemental and not as determinative evidence to justify the EPA's actions.[7] As Train correctly noted, the CHESS data had not been used to set air quality standards for SO_2. Thus even if the data were faulty, there would be no reason to challenge the existing standards, which had been set years before.

The hearing had calmed fears about Finklea's behavior but not about all of the results of the project itself. As a consequence, members of the House Sub-committee on the Environment and the Atmosphere decided that a compre-hensive scientific evaluation of the studies would be in order. When the appraisal was completed in November 1976, it signaled a low point in the EPA's brief history. A report over a hundred pages long severely criticized the CHESS studies and portrayed them as an example of poorly conducted research. Ac-cording to the scientific reviewers, "technical errors in measurement, unresolved problems in statistical analysis, and inconsistency in data" made the CHESS studies "useless" in determining the precise level of air pollution that caused harm.[8] As an illustration, the investigators cited their finding that improperly calibrated monitors had caused actual levels of SO_2 to be underestimated by as much as 100 percent.

Although it would be impossible to remedy retroactively these problems, the reviewers suggested that the EPA issue a formal notice listing the limitations of the research. Finally they disagreed with Train's assessment of how the CHESS material had been used. In contrast to Train's earlier testimony, the reviewers said that the data had been used to support the EPA's regulatory activities. In view of the studies' deficiencies, the evaluators concluded that the data should not be used to support regulatory decisions without explicit qualification.

The fault for all the problems encountered in the project was not entirely that of the researchers. During the course of their work, they were continually "pressured by EPA management-imposed time constraints to meet legislated mandates for promulgating new standards, hampered by inadequate mechanisms to detect and correct technical problems, and handicapped by budgetary and management restrictions placed on the program after it was well underway."[9]

The review panel observed that the EPA had known of many of the problems, but this awareness was not coupled with a willingness to correct them. To the distress of the reviewers, they found that the EPA was generally apathetic about remedying its shortcomings. Furthermore when the report was released, the agency's assistant administrator for research and development did not dispute the specific criticisms.[10] A few months later, during budgetary hearings on EPA research programs, John Quarles, who was then the agency's acting administrator, also declined to criticize the review.[11]

The severity of the review and the EPA's apparent willingness to accept the criticisms had several important implications. First, the agency found itself facing renewed suspicion about the quality of its scientific research. Such suspicion also encouraged polluters to question the validity of standards in other areas. Second, since the CHESS studies had been used to alert the public about the possible effects of sulfates, the congressional review of the studies offered a possible further explanation for the EPA's reluctance to establish air quality standards for sulfates.

Third, the CHESS studies' unfavorable reception emphasized the need for more rather than less research on sulfur oxides. Further research would be needed to determine whether sulfates are actually harmful at the levels the EPA had suggested. Since the studies had also questioned the adequacy of the existing SO_2 standards, more research would also be needed to examine their validity. Despite these needs, the experience with the CHESS studies might make any congressman reluctant to authorize large sums of money for another long-term epidemiological study of air pollution, especially if the EPA had responsibility for it.

Fourth, the attacks on Finklea made the EPA a less desirable work environment for many scientists. No one likes to have the news media question his honesty, and the possibility of similar attacks in the future is not an appealing inducement to conduct research for an embattled regulatory agency. In addition, the entire experience with the CHESS studies increased the likelihood that future researchers might tailor their findings to prevent another confrontation. In short, EPA's conduct of the CHESS project had produced few favorable outcomes and many of questionable value.

External Review of the Air Quality Criteria

During congressional hearings in 1974, many industries had complained that the EPA had not reviewed the air quality criteria as many people believed it should. The EPA responded by referring all of the criteria documents to an independent review panel, the National Air Quality Criteria Advisory Committee. When the committee finished its work in June 1976, its members recommended that all criteria should be the subject of major evaluations and revisions.[12] Revisions were needed, the panel said, because the existing criteria were outdated and did not adequately discuss such areas as the chemical transformation of pollutants. In a separate report on the sulfur oxides criteria, the committee stressed that they too were in serious need of revision. Not only had more information become available on SO_2, but on sulfates and sulfuric acid as well. Instead of a simple revision of the sulfur oxides criteria, the panel suggested that a combined document discussing sulfur oxides and particulates should be issued.

The EPA agreed and said that a combined and revised draft criteria would be available by August 1979, more than three years after the advisory committee's report and more than ten years after the document's initial publication. Thus what was judged to be an inadequate document would remain that way for several more years. This obviously displeased those opposed to the SO_2 standards, who thought that revision would also mean relaxation.

The Clean Air Amendments of 1976

When the members of the environmental subcommittees in the House and Senate conducted hearings in early 1975, their purpose had been to evaluate the need for changes in the Clean Air Amendments of 1970. Soon after the hearings were completed, the members of the subcommittees started to debate specific revisions. Unlike the near unanimity that had prevailed in 1970, discussion in 1975 was much more protracted and acerbic. That agreements were hard to reach is best illustrated by the large number of mark-up sessions in the Senate Subcommittee on Environmental Pollution and the House Subcommittee on Health and the Environment. In the former, twenty-four sessions were held between mid-June and November 1975. In the latter, members discussed dozens of separate proposals in their sixty-three mark-up sessions over a seven-month period.

Approval of the legislative changes in the two subcommittees was only the first step because their parent committees would also have to ratify the changes before sending them to the floor of the House and Senate. Thus when the Senate Committee in Public Works had finished more than a score of additional mark-up sessions and the House Commerce Committee nearly the same number, the committees reported out their legislative proposals in early 1976. That nearly a year had passed since the end of the hearings demonstrates the difficulty the committees had in reaching agreement. The lengthy period also demonstrates the need to devote some attention to what happened in both chambers.

Action in the Senate

While members of the Committee on Public Works heard many requests for changes in the existing law, most of the members were generally willing to leave unchanged the provisions of the 1970 amendments affecting stationary sources. In its proposal, the Senate committee did not recommend any changes in the boundaries for air quality control regions, the procedures for publishing air quality criteria, or the methods for proposing and publishing air quality standards. Some consideration had been given to the idea of requiring revision of the air quality standards, but the idea was rejected. Instead members of the committee suggested that the EPA revise the secondary standards for SO_2 because of recent information on sulfates.[13] In spite of this suggestion, the committee members made clear that the responsibility for such a revision would be the EPA's and not that of an environmental review board or some other group.

Agreement with an EPA position came with the committee's rejection of intermittent controls in favor of continuous emission controls such as scrubbers. An attempt to approve the use of intermittent controls by all polluters had been made in the environmental subcommittee, but the proposal was narrowly defeated. This left a bitter feeling with some members, who felt that scrubbers represented a costly and unproven technology that should not be forced on utilities. Even if the supporters of intermittent controls had won their case in the Public Works Committee, however, they still would have had problems. Most members of the House Commerce Committee were adamantly opposed to intermittent controls.

Because the Energy Supply and Environmental Coordination Act of 1974 had been considered and passed in such a short time, the Public Works Committee also directed its attention to the provisions of the energy law that affected the clean air legislation. Two important changes were proposed. First, because the provision was said to be "no longer necessary or appropriate," in view of the proposed legislative changes, the committee wanted to delete the EPA's authority to prevent conversions to coal if such conversions might lead to a significant risk to public health.[14] The EPA had interpreted this provision to apply only to sulfates. Second, the committee proposed that any emission source that had converted to coal as a primary source of energy need not comply with a state's emission limitations until January 1, 1979 (an extension of one year) or three years after the date of conversion, but under no circumstances later than mid-1980. The extension of the compliance date would not affect polluters' needs to limit their emissions so that primary air quality standards could be achieved. The House later accepted both of these changes.

For control regions not meeting the original deadlines for achieving air quality standards (mid-1975, or mid-1977 in some cases), the Senate extended the final compliance date until the end of 1978. Under certain conditions, this deadline could be extended further. Polluters that did not meet the new deadline would be subject to a monetary penalty equal to the cost of compliance. The purpose of the penalty would be to provide a definite incentive to reduce pollution. The provision on the compliance date extension was another with which the House eventually agreed.

To this point, the Public Works Committee's attitude on how the Clean Air Amendments should be changed reflected general satisfaction with the existing procedures and a heavy dose of incremental decision making. The broad outlines of the amendments and many of the specific provisions as well would be left intact. In many respects, therefore, the proposal represented an effort to introduce mid-course corrections and to provide the EPA with further guidance.

One topic caused considerable rancor; it concerned what the Congress should do about nonsignificant deterioration (NSD). During the hearings and debates on the 1970 legislation, little attention had been devoted to what the concept meant or how it should be implemented. By 1973, however, the Supreme Court's decision upholding the doctrine ensured that NSD would be a highly charged issue.

The Supreme Court had ruled in favor of the Sierra Club and against the EPA's argument that air quality better than the national standards should be allowed to deteriorate. Both business and industrial groups immediately claimed that the Court had implicitly approved a no-growth policy in most areas, particularly large parts of the West. In response to the Court's action, the EPA proposed a new set of guidelines for NSD in August 1974. After a period for public comments and hearings, the EPA issued what it thought would be the final guidelines in December 1974.[15] They were intended to reduce the federal government's role in determining acceptable levels of deterioration; the states' governors would be allowed to determine the amount of deterioration that could occur in pristine areas.

Although the proposals reflected a desire to balance competing objectives, the strategy failed. Industrialists and environmentalists alike challenged the new guidelines in the courts as either inadequate, inappropriate, or in direct conflict with earlier court decisions. The EPA's position on NSD was equally unsettling to President Ford, who wanted to eliminate the concept altogether. Faced with several court challenges, the president's strong opposition, and its own uncertainty over how the Congress would react to the NSD regulations, the EPA delayed implementation in the hope that legislative guidance would be provided. In fact, only four days after President Ford expressed his desire to abolish the concept, EPA Administrator Russell Train wrote to Senator Randolph, the chairman of the Public Works Committee, telling him of the need for legislation that would provide a "reasonable degree of certainty."[16]

The Subcommittee on Environmental Pollution responded to Train's request, but in doing so it sparked an intractable controversy that was duplicated in the House Subcommittee on Health and the Environment. Most members of the Senate committee disagreed with the president's position and wanted to prevent significant deterioration. They were able to devise a plan that granted the states a great deal of flexibility to guide their own economic growth while minimizing increases in pollution.

The revised NSD strategy was generally acceptable to most members of the full Public Works Committee but not to some other senators. Once the committee reported the bill to the floor, the NSD section became the most hotly debated environmental issue of the congressional session. Major industries were particularly upset with the possibility of any NSD strategy. Electric utilities were concerned that such a strategy would prevent the construction of new coal-fired plants in the West. Coal companies feared that the strategy would be tantamount to national land-use planning, which, in turn, would prevent increased coal consumption and the expansion of strip-mining activities. The U.S. Chamber of Commerce, the most outspoken opponent, added that NSD would prohibit nearly all further growth in many states. It offered this appraisal: "Congress may be on the verge of outlawing future economic development—not only manufacturing but new housing, farming and recreation—across vast sections of the United States."[17] The accuracy of this claim generated debate among both supporters and opponents. For opponents, the choice was clearly defined. More stringent environmental constraints would be incompatible with conventional economic values. While the business community could tolerate some environmental regulations, NSD went too far because it told businesses where they could locate, what they could build, and how they could operate, or so they believed. In sum, many people thought it was incomprehensible that the Senate had such intentions.

The critics of NSD placed their hopes with several senators who tried to eliminate the NSD provisions or at least to delay their implementation for as long as possible. This group failed, and in the end there were too many senators in favor of the committee proposal.

Nearly fifteen months after the Subcommittee on Environmental Pollution had ended its hearings, the Senate passed its version of the Clean Air Amendments of 1976 by a vote of seventy-eight to thirteen on August 5. All efforts to weaken the bill had failed, but so had several amendments designed to strengthen it. The Senate's work was temporarily completed because there was nothing it could do until the House of Representatives acted.

Action in the House

In the House, consideration of changes in the clean air legislation closely paralleled action in the Senate. Significant deterioration and what to do about it received intense scrutiny. In its report on the legislation, for example, the House Commerce Committee devoted sixty-eight pages to the policy in contrast to the few indistinct references to NSD in the committee's report on the 1970 law. More important, the committee clarified the reasons why deterioration should be prevented. Unlike the Senate's report, which had glossed over the health considerations, the Commerce Committee emphasized that they represented a major justification for preventing the deterioration of clean air.

The committee noted that new information had come to light since 1971 "showing that the public's health is being harmed to some extent, perhaps seriously, even at levels below the national standards."[18] The idea that the standards might be inadequate did not indicate that all pollution should be prohibited or that the standards should be changed immediately, but it did mean that prudence could not be ignored. As the committee report put it, "The approach of unlimited air quality deterioration is particularly short-sighted at a time when all indications point to the likely necessity for tightening the ambient air quality standards to protect public health."[19]

A concern for health was a major reason for preventing degradation, but economic concerns were also cited. According to the committee, a policy of NSD would discourage industrial flight from heavily polluted areas to relatively clean areas, would protect states from the threat of environmental blackmail, and would also reduce the likelihood that one area would suffer economically because of the pollution caused in other areas.

A number of alternative strategies were considered before the committee finally selected its own preference, which was similar to the Senate's version. Once the House Commerce Committee's NSD proposal reached the floor, it was as controversial as the Senate's had been. Several amendments were offered to delete or delay the plan, but the efforts were defeated just as they had been in the Commerce Committee. Several proposals increasing the flexibility of states were approved, but the basic framework remained unchanged.

Although NSD was the most controversial issue involving stationary sources of pollution, there were other issues included in the Commerce Committee's version that were not discussed in the Senate proposal. Many of these items in the House bill were designed to have a profound effect on the EPA's procedures for defining acceptable levels of air quality.

Review of Standards. While members of the Senate had not felt the need for a mandatory review of the existing air quality standards, their counterparts in the House Commerce Committee disagreed. According to the congressmen, because of "the importance of the ... air quality standards, the continuing controversy over the standards, and the Committee's desire for continued independent scientific review of the [EPA's] exercise of judgment," the EPA was directed to complete a thorough review of the criteria and standards at regular two-year intervals.[20] In conducting this review, the EPA would have to rely on the assistance of a scientific review committee, which would be given complete independence in conducting its own assessment of the criteria and standards.

Basis of Administrative Standards. After a lengthy review of the court decisions involving EPA's authority to regulate leaded gasoline, the Commerce Committee decided that the agency's administrator needed more flexibility in his decisions on air quality standards. In response to this recognition, the committee proposed that the administrator be allowed to regulate pollutants, the emissions of "which in his judgment cause or contribute to air pollution which may reasonably be anticipated to endanger public health or welfare."[21] This phrasing represented a significant change from the existing law, which required the administrator to regulate pollutants that he judged to "have an adverse effect on public health and welfare."

The proposed revision would clearly be more liberal than the existing guideline, and the Commerce Committee explained why. First, the changed language would allow the administrator to ensure the availability of precautionary measures; conclusive proof of adverse effects would not have to precede regulation. Second, the administrator would be permitted to consider possible risks, to extrapolate from limited data, and to make reasonable predictions about likely trends. Third, the revision would serve as a tacit acknowledgment of the uncertainties and limitations of the available scientific data on the health effects of air pollutants. The committee emphasized its intentions when it said that the new language was necessary "to recognize inevitable limitations or inadequacies of the data base available to the Administrator, to prevent courts from restricting his health protective decisions to 'current data,' and to insure that courts will not circumvent Congress' statement of priority for preventive health protection."[22]

Agency Procedures and Judicial Review. At the same time that the committee liberalized the substantive requirements for setting standards, it stiffened the procedural requirements. Noting the existing law's failure to allow adequate opportunities to challenge the factual basis of proposed rules, the committee proposed that the EPA conduct public hearings on many of its rule-making activities, including the establishment of air quality standards. The requirement to hold hearings was not meant to restrict the EPA's discretion but rather to ensure that it would be tempered by what the committee called "thorough and careful procedural safeguards." Once a hearing began, witnesses could be examined in regard to "disputed issues of material fact . . . the truth or falsity of which [are] subject to evidentiary proof."[23] Apparently, then, an administrator's judgment that an air pollutant causes harm would not be subject to cross-examination, but the correctness of the underlying scientific data would be. Thus EPA scientists could find themselves defending the conduct of their research and the rationale for their conclusions. For scientists willing to work under these conditions, too cautious an interpretation of data could result.

One further procedural safeguard was added to the review process. Should one of the EPA's regulations or air quality standards be challenged in the courts, the courts would be authorized to determine whether substantial evidence existed to support the decision. The EPA opposed the idea because it believed that it might jeopardize earlier legal decisions based on the traditional rule that courts could invalidate regulations only if they were made in an arbitrary and capricious manner. A shift from the old rule to a test of substantial evidence would seem to transfer the burden of proof from the plaintiff to the EPA. This was not the committee's intent, however, because it specified that a plaintiff would still have the responsibility for showing that a rule or regulation was invalid.

Nonetheless the proposed shift would provide the courts with greater leeway in their decision making and increase the number of opportunities for plaintiffs to challenge environmental regulations. Both of these outcomes could deter the EPA from initiating potentially controversial actions.

Congressional Review. An amendment that was originally rejected in the Commerce Committee but was later adopted by the entire House created the possibility of a legislative veto of EPA regulations. According to the amendment, the House or the Senate could void any air pollution regulation if either chamber passed a resolution of disapproval within sixty days of the regulation's promulgation.

The use of such an arrangement had its origins in the Congress's desire to affect reorganizations in the executive branch. Eventually the Congress began extending the veto to cover substantive legislation as well. The proponents of such vetoes argue that many agencies abuse their discretion by establishing rules and regulations that are contrary to congressional intent. In the words of one congressman, no agency deserved the veto more than the EPA: "Their track record of arbitrary rules and regulations which have exceeded congressional intent is such that this legislative veto provision is needed and will make EPA more responsive to the public."[24]

The possible use of a congressional veto meant that scientific data might be the basis of an environmental regulation but that popular disapproval could cause its elimination. In other words, congressional opinion could determine that some regulations were undesirable, no matter how conclusive the data or how imperative the need for regulation.

Agreement in the Conference Committee

A week after the House passed the revised legislation by a vote of 324 to 68, the Conference Committee began its work on September 22, 1976. The disparities between the two versions were great, but after more than thirty hours in conference, agreement was reached on the following:

1. The Senate, which had not required the EPA to review the air quality criteria or standards, accepted the need to do so, but with the modification that the reviews be completed by the end of 1980 and at least every five years thereafter.
2. The Senate conferees accepted, without change, the House wording regarding the standards for administrative decision making.
3. At the Senate's insistence, the conferees deleted all references to a change in agency procedures and judicial review. Senator Muskie called the House proposal "perhaps the most onerous and potentially damaging amendment of all." According to Muskie, the proposal was an unmistakable effort to "mute EPA's aggressive regulatory voice and intimidate its personnel through extensive cross-examination and complicated rulemaking procedures."[25]
4. The congressional veto was modified to require a concurrent resolution of disapproval by both chambers instead of disapproval by a single chamber.
5. For purposes of NSD, additional increments of pollution would be allowed depending on the areas affected.

Among these provisions, only the one on NSD generated much conflict when the conferees reported their agreement to the House and Senate on September 30. Unfortunately for the bill's sponsors, this conflict was sufficient to prevent passage of the entire bill.

The Senate began debate on it as soon as the conference report was filed, but 1976 was an election year. Many senators were anxious to start serious campaigning, and adjournment was only a few days away. With only two days of debate possible, a few opponents could use a filibuster to kill the bill. This was precisely what Senator Jake Garn of Utah and a few of his colleagues decided to do because of their dislike for the bill's NSD policy. The chief sponsor of the legislation, Senator Muskie, tried to gather enough votes either to stop the filibuster or to delay adjournment, but he failed on both counts. The Senate voted to adjourn without taking a vote on the clean air bill. Over eighteen months of negotiation and compromise had been wasted. Further action on the legislation would have to wait until the beginning of the Ninety-fifth Congress in January 1977.

The Clean Air Amendments of 1977

The long debates in 1975 and 1976 had not produced a new law, but they did set the stage for action in 1977. Soon after the new congressional session began, clean air bills very similar to those passed in the House and the Senate in 1976 were introduced in the respective chambers. Soon thereafter, the Senate Subcommittee on Environmental Pollution held four days of hearings on the proposed legislation. Not unexpectedly, much of the testimony was remarkably predictable. Most industrialists opposed NSD and favored intermittent controls, a delay in meeting compliance deadlines, the need to balance environmental objectives with other social and economic goals, and the need for a thorough review or revision of the criteria and standards. A spokesman from the aluminum industry, for example, said that a fixed date for attaining the standards should be replaced with a program designed to encourage "reasonable progress" toward them. The American Iron and Steel Institute offered a similar idea when it recommended that air quality standards "be considered as goals to be approached as expeditiously as practicable."[26]

The Manufacturing Chemists Association, the National Association of Manufacturers, and the Edison Electric Institute all questioned the data base for the air quality standards. The National Coal Association had recommendations in nearly all areas. First, the association said that a thorough review should be conducted to examine the appropriateness of existing margins of safety, the feasibility of meeting air quality standards, and the validity of EPA's criteria documents and monitoring data. Second, a committee with members appointed by the president should review the standards. Third, because a policy on NSD was "too complex to finalize," action on it should also be deferred pending further study and review. The association's main point was that more information would be needed before further regulatory controls could be justified. The coal lobby concluded that their requests were both simple and reasonable.[27]

A few weeks after the Senate committee finished its hearings, most of the same groups appeared before the House Subcommittee on Health and the Environment. Many of these witnesses made the same points they had brought before the Senate committee. In fact, several people presented virtually identical testimony to both committees. One major difference, however, was the appearance of EPA's new administrator, Douglas Costle, whom President Jimmy Carter had recently appointed.

Unlike his predecessor, President Carter strongly favored environmental protection. Thus it was not surprising that Costle maintained that all facilities converting to coal should meet emission limitations sufficient to achieve the federal air quality standards from the time of conversion, instead of some future date. Costle also indicated the administration's support for a policy that prevented the significant deterioration of air quality in clean areas. "Protection of these areas," said Costle, "is too important to allow possible litigation to create any uncertainty over current EPA regulations."[28]

The Committees' Reports and Legislative Action

When the House Commerce Committee and the newly renamed Senate Committee on Environment and Public Works issued their reports on the Clean Air Amendments of 1977, the discussion of stationary sources was very similar to the respective reports of the preceding year. The similarity between the two proposals ensured the return of controversy once the committee bill reached the floor of the House. Several of the committee members had failed to soften the NSD provision within the Commerce Committee, but they did succeed when all the representatives had a chance to vote on it. After debate, the House approved, 237 to 172, an amendment that would allow a governor to let allowable pollution increments be exceeded eighteen days a year in certain relatively clean areas. Opponents of the amendment claimed that it would nullify the intent of the NSD policy.

In the Senate Public Works Committee, the debate over NSD was no less acrimonious, but the committee made only a few minor changes from its position in 1976. Once the proposal reached the Senate floor, several senators from western states tried, but failed, to weaken the NSD provisions.

Both the House and Senate overwhelmingly approved the Clean Air Amendments of 1977 in late spring, but the conferees did not begin their negotiations until late July. After more than a week of meetings, a bill acceptable to the conferees was sent to each chamber and quickly approved by both in early August.

The new law included many of the provisions that the conferees had accepted in 1976. Those that were the same or only marginally changed included the ones on coal conversions, standards for administrative decision making, and review of air quality criteria and standards. Revised implementation plans would be required from states that had not attained the original 1975 or 1977 compliance deadlines. For these nonattainment areas, ambient standards for SO_2 would have to be achieved no later than the end of 1982. In the interim, reasonable progress would be required. The Senate conferees had again balked at efforts to introduce major modifications in the EPA's rule-making procedures, but they had accepted the House proposal after making several important changes. Among these were two that deleted the "substantial evidence" test and the opportunity to cross-examine witnesses at EPA's public hearings. The senators had also been successful in their efforts to eliminate the House proposal allowing the Congress to veto any of the EPA's clean air rules or regulations. Finally the policy on NSD was different in that it gave the states even more control over implementation and included a modified version of the weakening amendment, which the House had adopted.

When President Carter approved the Clean Air Amendments of 1977 (Public Law 95-95), on August 7, he signed a detailed and highly complex bill. The new law was over 180 pages in length—three times as long as the 1970 law and six times as long as the 1967 Air Quality Act. In several ways, the new law did not include any major departures from the one it amended. It did, however, diminish the EPA's discretion and flexibility. The EPA's responsibilities for maintaining pristine areas were cut substantially in favor of a larger role for the states. The many attempts to guide and control the agency's policy making also reflected growing and widespread dissatisfaction with the consequences of social regulation. Proposals for legislative vetoes, stringent procedural requirements, and independent review committees reflected the belief of many congressmen that the EPA had become too powerful because of too many unlicensed excesses.

The new law also created a different order of priorities. According to the 1970 version, protection of the public's health and welfare was a paramount value, and polluters were expected to comply with the law's mandate or to cease their operations. By 1977, however, few polluters had had to make this choice, and a reordering of priorities was underway. In the words of the Senate report, the law attempted "to balance the economic aspirations of the country with the need for protection of the public health and welfare from the adverse impacts of air pollution."[29] The House report was equally straightforward when it noted that one purpose of the new law would be "to provide express recognition of the economic and energy factors which should be considered."[30] In sum, while both the House and the Senate reemphasized the need to protect the public's health, they signaled that the achievement of acceptable levels of air quality would depend on their compatibility with other social values.

In summary, the new law did not resolve what has become a recurring question: what should be the relative importance of social, economic, and environmental values in determining acceptable levels of air quality? This question was answered one way in 1970, but the tenacity of regulated groups and changes in the nation's policies and priorities ensured that the debate would continue.

Implementing the New Law

One of the purposes in passing the Clean Air Amendments of 1977 had been to clarify Congress's intent about what the EPA should or could do to reduce air pollution. In spite of the extensive detail with which the Congress outlined the EPA's duties, the new law did not simplify the agency's problems. For groups disappointed with congressional action, the year after the law's passage provided still another opportunity to influence its implementation. Several events illustrate this point and the fact that the debate about acceptable levels of air quality is unlikely to end.

First, soon after the EPA issued proposed rules for the implementation of the new policy on NSD, industrial and environmental groups attacked the rules as contrary to congressional intent. Other environmental groups ignored the opportunity to speak at public hearings on NSD but challenged the rules in several court suits. The Department of Energy also asked the EPA to modify its rules.

Second, a government task force reported in early 1978 that large increases in coal use could cause undesirable health and environmental effects.[31] A few months later, a probusiness group claimed that certain environmental rules had to be relaxed in order to avoid a catastrophic energy crisis that could cause massive unemployment and a severely weakened economic system.[32] Air quality regulations were high on the list of rules to be changed.

Third, after its announcement that revised draft criteria for sulfur oxides would be issued in August 1979, the EPA amended its statement. The draft would be issued in May 1980, with final publication scheduled for December 1980, the last possible month under the requirements of the clean air law. The EPA also said that research designed to provide information for a sulfate standard could not be completed by 1980 or 1981 as originally expected. As a result, a standard for sulfates might not be set until 1983 or later, despite the added flexibility that the Clean Air Amendments of 1977 had provided. This was paradoxical because Douglas Costle, the EPA administrator, announced in late 1978 that his agency would be willing to set certain environmental standards without perfect information.[33]

The new criteria and standards for SO_2 are not likely to be free from criticism. Unless the standards are relaxed, polluters will repeat their earlier complaints and claim that the scientific data are inappropriate or misinterpreted. Polluters have an economic interest in the outcome of the argument, but others without this interest continue to express concern over the validity of standards for sulfur oxides. At a medical symposium conducted in the spring of 1978, for example, several speakers challenged the existing standards for SO_2, and one scientific researcher suggested their elimination.[34] Another said that no scientific data existed to implicate sulfates as a cause of harm.

Fourth, the Carter administration's support for environmental protection appeared to dim as the rate of inflation remained at high levels. By the summer of 1978, the newly created Regulatory Analysis Review Group, which the president had authorized, had asked the EPA to soften several rules designed to implement the Clean Air Amendments. The *New York Times* reported that the anti-inflation drive, though only a few months old, had already had a "chilling effect" on the bureaucrats who were writing environmental regulations.[35] By mid-November, the White House had suggested the establishment of still another panel, whose major purpose would be to review and coordinate proposed health, safety, and environmental regulations to ensure that they are not too costly. Some regulatory officials anticipated that the review "could lead to politically or economically motivated delays or cutbacks" in their efforts.[36]

Each of these situations is hardly new. Indeed the themes are variations of past discussions that are likely to be repeated many times in the future; after more than two decades of debate, certainty over the acceptability of various levels of sulfur oxides is still absent.

Notes

1. EPA, Science Advisory Board, "Review of the CHESS Program: Executive Summary" (March 14, 1975), p. 3.

2. W.B. Rood, "EPA Study—The Findings Got Distorted," *Los Angeles Times,* February 29, 1976, p. 1.

3. *Coal News,* March 12, 1976.

4. *Congressional Record,* daily edition, March 18, 1976, p. H2107.

5. U.S. House of Representatives, Committee on Science and Technology, Subcommittee on the Environment and the Atmosphere and Committee on Interstate and Foreign Commerce, Subcommittee on Health and the Environment, joint hearing, *The Conduct of the EPA's "Community Health and Environmental Surveillance System" (CHESS) Studies,* 94th Cong., 2d sess., 1976, p. 59.

6. Ibid., p. 71. Shortly after completion of the draft version of the CHESS report, Finklea became director of EPA's National Environmental Research Center. At the time of the hearings, he was director of the National Institute for Occupational Safety and Health.

7. Ibid., p. 85.

8. U.S. House of Representatives, Committee on Science and Technology, Subcommittee on Special Studies, Investigations and Oversight and Subcommittee on the Environment and the Atmosphere, Committee Print, *The Environmental Protection Agency's Research Program,* 94th Cong., 2d sess., 1976, p. 12.

9. Ibid., p. 11.

10. *Washington Post,* November 25, 1976, p. A16.

11. U.S. House of Representatives, Committee on Science and Technology, Subcommittee on the Environment and the Atmosphere, hearings, *1978 Authorization for the Office of Research and Development, Environmental Protection Agency,* 95th Cong., 1st sess., 1977, p. 9.

12. EPA, Science Advisory Board, National Air Quality Criteria Advisory Committee, "Report on Air Quality Criteria: General Comments and Recommendations" (June 1976), p. 1.

13. U.S. Senate, Committee on Public Works, Report No. 94-717, *Clean Air Amendments of 1976*, 94th Cong., 2d sess., 1976, p. 8.

14. Ibid., p. 47.

15. *Federal Register* 39 (December 5, 1974):42510.

16. John C. Whitaker, *Striking a Balance* (Washington, D.C.: American Enterprise Institute, 1976), p. 109.

17. *Congressional Quarterly Weekly Report,* May 1, 1976, p. 1035.

18. U.S. House of Representatives, Committee on Interstate and Foreign Commerce, Report No. 94-1175, *Clean Air Act Amendments of 1976,* 94th Cong., 2d sess., 1976, p. 85.

19. Ibid., p. 107.

20. Ibid., p. 155.

21. Ibid., p. 33.

22. Ibid., pp. 34–35.

23. Ibid., p. 262.

24. *Congressional Record,* daily edition, September 15, 1976, p. H10119.

25. Ibid., daily edition, October 1, 1976, p. S17535.

26. U.S. Senate, Committee on Environment and Public Works, Subcommittee on Environmental Pollution, hearings, *Clean Air Act Amendments of 1977, Part 2,* 95th Cong., 1st sess., 1977, pp. 493, 1015.

27. Ibid., p. 467.

28. U.S. House of Representatives, Committee on Interstate and Foreign Commerce, hearings, *Clean Air Act Amendments of 1977, Part 2,* 95th Cong., 1st sess., 1977, pp. 1679, 1681.

29. U.S. Senate, Commiteee on Environment and Public Works, Report No. 95-127, *Clean Air Amendments of 1977,* 95th Cong., 1st sess., 1977, p. 2.

30. U.S. House of Representatives, Committee on Interstate and Foreign Commerce, Report No. 95-294, *Clean Air Act Amendments of 1977,* 95th Cong., 1st sess., 1977, p. 2.

31. *Federal Register* 43 (January 16, 1978):2229.

32. The Heritage Foundation's report, "Closing the Nuclear Option: Scenarios for Society Change," is cited by *Environment Reporter: Current Developments* 8 (April 28, 1978):2049-2050.

33. Ibid., 9 (September 15, 1978):914.

34. Ibid., 8 (March 31, 1978):1862-1864.

35. *New York Times,* June 12, 1978, p. 1.

36. *Washington Post,* November 18, 1978, p. A4.

10 Evaluating the Process

Many battles have been fought over the definition of acceptable levels of sulfur oxides. In most instances, the outcome of these battles has been many controversial political and scientific decisions. Some groups have strenuously argued that the decisions are without merit, and others have been equally vociferous in stressing that the decisions demonstrate an inadequate concern for the public health. The probable consequences of improper decisions emphasize the importance of this debate and the need to evaluate the entire process, both from a narrow health-related perspective and from a larger societal perspective.

Because the determination of environmental standards reflects a complex set of value judgments, the decisions cannot be measured against objective criteria to indicate whether they are right or wrong. Accordingly it is impossible to assess directly the soundness of these decisions. Although this may be the case, it does not mean that indirect measures are absent. For example, the quality and quantity of the EPA's scientific research can serve as indicators of the quality and soundness of the agency's regulatory decisions. Thus examination of such things as the overall management of EPA's research programs and the level of resources devoted to research on air pollution can provide some measure of the merits, defensibility, and appropriateness of the agency's judgments on sulfur oxides and other air pollutants.

Adequacy of Resources

Has sufficient money and manpower been provided to conduct the research needed to set appropriate environmental regulations? Each clean air law has authorized particular amounts of funding for the federal government's environmental agencies. The initial law, passed in 1955, authorized $5 million per year. The Clean Air Amendments of 1977 authorized more than $500 million per year. Despite the huge increase in authorizations, congressional appropriations have rarely matched the amount supposedly needed to implement the laws. In fact, the last time that appropriations matched or exceeded authorizations was in fiscal year 1964, when Congress provided $13 million, $3 million more than had been authorized. Between 1964 and 1967, HEW's environmental agencies were given most of what the Clean Air Act of 1963 authorized, but since 1968, appropriations have not kept pace with authorizations. By fiscal year

1975, actual appropriations amounted to slightly more than one-third of the legislative authorization.

The EPA has never received as much as the Congress has authorized, but Congress is not entirely to blame. The OMB, which must approve all budgetary requests before they are sent to the Congress, has never allowed the EPA to ask for full funding of its air quality programs. For fiscal year 1973, as an illustration, the statutory authorization was $475 million, but OMB permitted the EPA to ask for only $160 million. More important, the OMB has not allowed EPA's budget requests to keep pace with inflationary price increases. The amount that the OMB approved for EPA for fiscal year 1976 was actually 12 percent below the 1973 level. The situation improved only marginally in the following two years.

The reasons for EPA's lack of financial success are not hard to find. Congress is notorious for authorizing far more than it is willing to spend. Hefty authorizations are frequently seen as a means of giving some people the impression of action and concern. Within the Congress, the EPA also faced the problem of having to deal with a staunch critic of its operations, Congressman Jamie Whitten (D.-Miss.). He chaired the House Appropriations Subcommittee with jurisdiction over the EPA from 1971 to 1974, and he was always willing to criticize the EPA or to find ways to divert its limited resources. Rarely did he create a hospitable environment for the EPA. The OMB's lack of generosity reflects both its desire to minimize deficit spending and presidents' opposition to or incomplete support for some environmental programs. Shortly before Russell Train resigned as administrator of the EPA in early 1977, for example, he complained that the agency had not been given enough money to fulfill its responsibilities under the clean air legislation.[1]

The relative paucity of funds for implementation of the legislation also helps to explain why the EPA has not been able to develop an adequately funded research program on air pollutants. Each time the Congress appropriates funds, the EPA must divide these funds among such competing areas as research, monitoring, enforcement, and the development of control technologies. Few areas are likely to get as much money as their managers need or can use. In the areas of research on the health effects of air pollutants, however, the situation seems to be especially critical. An EPA task force estimated in 1973 that a minimally adequate research program on the health effects of air pollution would cost at least $25 million a year, and an optimal program would cost another $15 million.[2] A year later, one EPA official said that $50 million per year would be the amount needed to conduct an adequate research program. These amounts are far less than the EPA has actually spent to study the health effects of all air pollutants (table 10-1).

At the beginning of the Carter administration, many people believed that a larger commitment to environmental protection would result. To a certain extent, they were correct. When President Carter took office, he increased the

Table 10-1

EPA Budgets for Research on the Health Effects of Air Pollutants

(millions of dollars)

Fiscal Year	Request to Congress	Appropriation	Amount Obligated
1972	N.A.	5.5	5.4
1973	N.A.	12.4	10.7
1974	12.7	12.5	10.9
1975	20.3	20.3	14.2
1976	20.8	22.8	13.2
1977	19.2	19.8	20.9
1978	15.8	19.1	18.0

Source: EPA, Office of Planning and Management.

EPA's budget, but little of the extra money was intended for research on air pollutants. In the first budget for which Carter had full responsibility (fiscal year 1979), he requested nearly a 12 percent increase in expenditures for EPA's regulatory activities, but a decrease in spending for research on air pollutants and ways to control them.[3]

For several years, the EPA did not prepare annual estimates of the resources necessary to establish an adequate research program on air pollutants. Even if it had, it is doubtful whether they would have been of much use. One of John Finklea's major complaints about the EPA's budget managers was that they frequently made decisions on how to allocate resources without the benefit of technical expertise.[4] When the General Accounting Office (GAO) reviewed the EPA's research programs on air pollution, it found a poorly managed budgetary system, with some budget managers unable to indicate how much they were spending on certain projects.[5]

If a lack of financial resources is one reason for the existence of an inadequate research program, a related problem concerns the availability of properly trained researchers. Capable researchers are essential to any agency, but especially to the EPA, which usually finds its regulations under intense scrutiny. Reputable scientists can provide useful data and can help to ensure the acceptance of regulations based on these data.

Well-trained scientists are essential to credible research, but the number of scientists who are qualified and willing to conduct research on air pollution is exceedingly small. Unlike the medical profession as a whole, where some surpluses are evident, epidemiologists trained to do research on air pollution are few in number. Even if sufficient numbers of researchers could be found, it is likely that the EPA would have trouble recruiting or keeping them once hired.

Many scientists are reluctant to work in a situation where research objectives frequently change, where funding is scarce, and where researchers often find themselves under bitter attack. The opportunities for higher pay in private enterprise are also appealing to many scientists.

Despite these drawbacks, many highly qualified scientists work for the EPA. Unfortunately for the agency, the turnover among them has been a serious problem. During one two-year period, for example, the CHESS studies had three directors; similar turnover rates are not unusual in other research programs.[6] High rates of turnover can disrupt the continuity of research projects and often delay them while new directors familiarize themselves with the work and their new subordinates.

Limits on the EPA's money and manpower have often contributed to low morale among researchers. Similarly many projects that researchers consider imperative must be delayed or reduced in scope. The EPA's dependence on annual appropriations has also meant that some important long-term projects are not scheduled, while others have their budgets unexpectedly reduced in mid-course, which happened with the CHESS studies. Budget cuts similar to those experienced by the CHESS researchers have not been uncommon. According to one Senate investigation, EPA's scientists have been reluctant to plan or schedule certain research because long-term funding could never be assured.[7]

The Management of EPA's Research Programs

Adequate amounts of money and skilled researchers are critical determinants of credible research, but equally important is the manner in which policy makers handle the resources they do have. Careful use of resources can overcome many problems, while mismanagement often leads to results of dubious value. Policy makers must appreciate the problems of scientific research and establish an organizational structure that effectively incorporates scientific information. Policy makers without technical backgrounds must also establish a close rapport with scientists so that the latter can communicate the limits and significance of their findings.

These seem to be reasonable requirements, but the constraints evident in any bureaucratic agency can make them difficult to achieve. For a regulatory agency faced with scarce resources, unlimited critics, and statutory deadlines, the task may be nearly impossible. Two examples demonstrate this point.

First, as in most other governmental agencies, the tenure of key decision makers is often shorter than the time frame needed to resolve some problems. As the discussion on SO_2 and sulfates has shown, the research process is costly, complicated, long term, and challenging. Years can pass before useful results are available, and new findings can indicate the need for additional investigation. Because of these complications, policy makers with short tenures often do not appreciate the limitations of scientific research. As Edwin Hargrove has pointed out, many policy makers are interested in making a personal mark in the short run. They thus emphasize quick achievements to the detriment of problems that require extended attention.[8]

The EPA has been especially prone to the problems of policy makers with short-term perspectives. Three different people served as administrators and two others as acting administrators in the EPA's first seven years. Of these five, the first two, William Ruckelshaus and Russell Train, spent much of their time testifying before congressional committees and defending environmental regulations. Neither had much time to become involved in extensive long-range planning. At lower levels, the problems have been even more acute. When the EPA was originally proposed, Congress was told that the new agency would be independent, but the Nixon administration regularly used political appointees to ensure that this would not be the case. The White House accomplished this objective by filling many key positions in the EPA with people selected on the basis of their political credentials instead of their environmental expertise. At one time, the EPA had more top-level political appointees than the combined total of four other regulatory agencies.[9]

In addition to the political appointees, who were probably more interested in the president's fortunes than in the EPA's, the agency regularly experiences exceptionally high rates of turnover among its personnel, which contributes further to short-term perspectives. In one recent year, for example, more than 30 percent of the EPA's staff resigned, and the agency's average turnover rate is substantially higher than the average for all federal agencies.[10]

Second, as a regulatory agency, the EPA's success is often measured by the number of enforcement actions it brings and not in the amount of research it conducts. Enforcement actions are viewed as a positive sign of accomplishment, while research by itself does not create a clean environment. Many militant environmentalists share this perception and exert considerable pressure on the agency to emphasize enforcement at the expense of research. Thus policy makers with short-term perspectives may believe that enforcement is more important than research. A former assistant administrator for research and development described the problem well: "From the regulatory side of this conflict emerges a resentment of the need to 'waste' funds on nonregulatory activities and a frustration over the inability of the research activity to provide the needed and required information on demand."[11]

Another disadvantage of being a regulatory agency is Congress's frequent inclusion of arbitrary deadlines in environmental laws. Congressmen may think that statutory deadlines demonstrate their good intentions, but time limits can create enormous problems. They often require the EPA to set standards within a particular time period, regardless of the availability of scientific data or of the length of time required to conduct an adequate assessment of the situation. The problem was especially evident in the early 1970s, when Congress passed many environmental laws that required almost immediate establishment of environmental standards. After being promulgated in haste, many of these standards have not been changed. Should the EPA fail to act on time, the Congress has included many opportunities for courts to enforce deadlines.

Deadlines ensure that some action is taken, but the outcome can be regulations based on incomplete or poorly analyzed data. Indeed if policy makers are told that appropriate research will take longer than Congress has allowed, the research may not be conducted at all, and serious gaps in knowledge remain unfilled. For research that is started, there are many incentives to produce quick results, and this is especially true for an agency expected to react as quickly as possible to nearly all environmental problems. When the Kennecott Copper Corporation challenged the SO_2 standard in 1971, for example, many scientists were taken away from other projects in order to gather and analyze data that could be used to buttress the EPA's case.

A further problem that faces the EPA is the poor relationship between its policy makers and their use of scientific data. Good scientific data are wasted when they are poorly used or not used at all. Past EPA decision making has exhibited both tendencies. As an illustration, many of the agency's early policies on nonsignificant deterioration apparently were formulated without scientific input.[12] In another case, a study of the EPA's research planning found that the agency's scientists often do not have adequate opportunities to review the scientific credibility of new regulations.[13] Part of the trouble lies in the composition of EPA's leadership. While most of the agency's decisions are heavily dependent on scientific research, lawyers tend to be found in many of its key positions. Lawyers are well suited for many tasks, but most do not have the competency to appraise the adequacy or quality of complex scientific research. This causes problems with researchers, and the frequent result is resentment on both sides.

The difficulties in establishing good relations between researchers and policy makers are compounded because of the large number of scientific disciplines required to support environmental regulations. Perhaps no other agency confronts policy problems that demand so much scientific information from so many disciplines, including botany, chemistry, biology, statistics, engineering, and health sciences.

In view of all these problems it is not surprising that many investigations have found that the EPA is sorely deficient in its management of research. The authors of the GAO study referred to earlier said that they were unable to discern a clear pattern of priorities for research on air pollutants.[14] As a result, EPA scientists often have little idea of what is expected of them. One congressional study asserted that the need to respond to changing priorities created a "pollutant-of-the-month" syndrome.[15] Still another report criticized the EPA for its undue emphasis on pet projects, which often bear little relation to the agency's overall research requirements.[16]

These evaluations of the EPA's research programs do not represent isolated criticisms. Indeed by the mid-1970s, the problems of organizing and managing the EPA's research had become so noticeable that they had attracted the attention of the GAO, the National Academy of Sciences, and committees in both the House and the Senate. Each of these reviews was highly unfavorable, and the EPA reorganized its research operations in 1975 in an effort to improve their efficiency. The initial response to the reorganization was positive, but by 1977, another external review stressed that poorly planned research still prevailed. "EPA's research and development programs alone," observed the National Academy of Sciences, "are not adequate to serve either its current requirements for technical data and analyses or its long-term need to expand knowledge of the physical, biological, economic, and social phenomena related to problems of environmental regulation."[17]

This discussion presents a bleak picture of research in the EPA. So many constraints and problems exist that they continually bring into question the reliability of the agency's scientific data. To a large extent, the problems explain polluters' resistance to environmental regulations and predict recurring controversy in the future when new standards are proposed.

The problems are soluble, but not in the foreseeable future. The research problems are large, and many are not easily solved. The EPA can improve its scientific work, and in many areas it has. Nonetheless many other problems are outside of its control—limited resources, uncertain congressional guidance, and a natural reluctance among polluters to trust a regulatory agency.

It has often been suggested that it is inappropriate for a regulatory agency to conduct research to support its regulations. This was one reason that the Atomic Energy Commission was divided into the Nuclear Regulatory Commission and the Energy Research and Development Administration. Regulatory agencies are claimed to have a natural bias to select scientific data that support their regulations while discounting other data. Advocates of change also claim that separate institutes could eliminate conflicts of interest and protect governmental scientists from a loss of objectivity by their involvement in the political process.

It can be difficult to detect bias, but even when it is unintended, its appearance would seem to be the likeliest in instances in which controversial decisions have already been made or where a regulatory agency is under attack from regulated groups.[18] These traits seem to describe exactly the situation with the EPA, SO_2, and sulfates. During the congressional hearing on the CHESS studies, for example, several witnesses asked whether it is appropriate for the EPA to sponsor such research. In the words of one scientist, "regulatory pressures and the perceived need to have some numbers for use by decisionmakers" probably contributed to overinterpretation of the results.[19] In short, even if the EPA improved its performance, its data might still lack sufficient credibility to discourage a presumption of institutional bias.[20]

Not all of the blame for EPA's difficulties is the agency's. The EPA has operated with limited political support in an exceedingly hostile political environment. It has tried to implement many policies that Congress has not clearly formulated. In some instances, in fact, Congress has not understood the implications of its directives to the EPA. The continuing conflict over nonsignificant deterioration provides an excellent example. It may also be the case that the Congress has asked the EPA to do too much. Not only is the EPA the largest regulatory agency in the federal government, but it probably has more statutes governing its operations than any other. While the Congress has been willing to increase the EPA's operational responsibilities, it has not matched these responsibilities with adequate resources, legislative support, or consistent direction.

These characteristics would seem to predict that the EPA has been a dismal failure. There are surely enough reasons to justify any failures that occur. Yet however compelling the predictions, they are in error. Substantial reductions in air pollution have occurred since 1970, and further reductions are likely if the Clean Air Amendments of 1977 are successfully implemented. Unfortunately, however, as the continuing uncertainty over the appropriateness of the SO_2 standards has shown, their achievement will not necessarily mean that the public's health is protected from the effects of polluted air.

At least one caveat is in order: levels of pollution have decreased in certain areas in recent years, but the gains could be eliminated because of the nation's energy situation. Efforts to increase substantially the use of coal will place enormous pressure on the EPA to relax enforcement actions and, perhaps, its standards for nonsignificant deterioration. The result will be higher levels of pollution and more debate over the air quality standards.

The Implications of Social Regulation

In chapter 1, efforts to protect the environment were described as fitting into a general pattern of regulatory policy in which Congress passes ambiguous laws and then delegates to administrative agencies the responsibility for formulating specific regulations. According to this model, knowledge that a policy is regulative assists one in understanding the political processes that surround the regulation. The government's efforts to set acceptable levels of sulfur oxides demonstrate the utility of this model. At the same time, however, these same events reveal that environmental protection represents a special form of social regulation. In this form, policy makers become thoroughly involved with the daily operations of regulated groups. The involvement of such social regulators as EPA, the Occupational Safety and Health Administration (OSHA), and the Consumer Product Safety Commission extends as far as telling different industries how they can produce their goods rather than just defining how these items can be sold, which is usually the case with traditional forms of economic regulation.

Since social regulation is so distinct from its economic counterpart in the mode of regulation, the question arises whether any other differences exist. This study of the definition of acceptable levels of air quality points to several that deserve attention. Perhaps the most noticeable difference between the two forms of regulation involves the pattern of relations that exists between regulators and regulated groups. In his seminal work, *The Symbolic Uses of Politics,* Murray Edelman cites the conventional wisdom that regulation is pursued only when it is acceptable to the interests of the regulated.[21] More often than not, regulated groups actually encourage regulation for its symbolic value as much as for its assurance of stability and economic well-being. Additionally the conventional wisdom indicates that changes in economic regulations will occur incrementally and will be the subject of negotiation to ensure their acceptability to and marginal impact on regulated groups.

Although the symbolism is retained, quite the opposite pattern exists for social regulators, which have the power to upset the status quo and to require the expenditure of billions of dollars on economically nonproductive investments. In the case of environmental protection, because air and water have traditionally been used as free resources, environmental regulations almost always require major changes in well-established practices. As a consequence, social regulators are likely to create a great deal of instability and economic insecurity among regulated groups. In turn, they generate apprehension and intense opposition to regulation instead of the arrangement that Edelman and many others have hypothesized. This opposition is heightened because changes in social regulations are not necessarily incremental or dependent on negotiation. In fact, the reliance of social regulators on scientific and technical data means that significant changes in regulations can occur over a very short period as new data become available or as new health and environmental hazards are discovered.

Although such changes are distinct possibilities, the experience with the case of sulfur oxides has demonstrated that economic and political pressures can be sufficient to maintain reliance on incremental decision making. The standards that the EPA selected for SO_2 in 1971 are based on data collected no more recently than 1968. Since that time innumerable studies have produced new scientific findings indicating that SO_2 by itself probably does not cause adverse effects, though sulfates and SO_2 in combination with particulates probably do. Despite the growing frequency with which scientists document these conclusions, the original air quality standards remain unchanged. Surely this example demonstrates the supremacy of politics over science in this instance.

Another way in which social and economic regulation differ involves the number and kinds of groups that are subject to control. With the latter form, for example, a single regulatory agency, such as the Civil Aeronautics Board, tends to concern itself with a single industry or at least a group of closely related industries. Under these conditions, the success of the regulatory agency is so highly dependent on the industry that the agency often becomes a captive of regulated groups.[22] In short, both the agency and the industry benefit from the other's cooperation.

In contrast to this arrangement, social regulators tend to affect entire groups of diverse industries and to unite them in opposition. Many of these industries have little in common except their opposition to social regulations that they believe are too costly and restrictive. Their shared interests in opposing regulations far exceed any benefits that might accrue to them because of their cooperation with regulators.

Affecting many different industries also means that social regulators tend to alienate other governmental agencies that are organized around particular constituencies. Put in other words, social regulators cannot avoid overlapping jurisdictions with such agencies as the Departments of Labor, Energy, Commerce, Agriculture, and Transportation. As the discussion of sulfur oxides has demonstrated, the federal government's environmental agencies have faced constant opposition from other governmental units that are trying to protect their constituents' interests. These other agencies have often been as recalcitrant as the regulated industries themselves.[23] The result is that progress is slowed and regulated industries are encouraged to continue their opposition to social regulation.

Still another difference between social and economic regulators lies in their relative dependence on affected groups for the information necessary to develop regulations. Economic regulators tend to be especially dependent on the willingness of businessmen to provide accurate data about costs, sales, profits, and levels of production. Without these data, it could be impossible to establish viable regulations. Such an arrangement works to the advantage of regulated groups because they can selectively filter the information provided and therefore influence the nature and scope of the regulations. The federal government's heavy reliance on data from the petroleum industry probably provides the best illustration of this situation.

In stark contrast, social regulators can generate many of their own data and virtually ignore those of industry or business. Since regulated groups probably do not know exactly how a social regulator's data are collected or analyzed, skepticism is increased and credibility is diminished. These problems are enhanced because the regulated groups have no control over the design of research or the selection of researchers.

Finally the conventional wisdom about regulatory agencies suggests that as they age, they lose their vitality and establish a cozy relation with regulated groups. When regulators reach maturity, this theory holds, their interests coincide with those of the industry, and a symbiotic relation results. Such coexistence may occur with economic regulation, but it is unlikely that it will happen for agencies involved in the protection of safety, health, or the environment. Instead of establishing harmonious relations with regulated groups, social regulations tend to exacerbate discord. In Paul Weaver's words, cooperative relations are unlikely because agencies like the EPA are created "to operate as the adversaries of the interests that they regulate, and typically it is as adversaries that they administer the law."[24]

The trends of social regulation suggest that the relations between the adversaries will probably become even more acrimonious in the future than they have been in the past, for several reasons. First, in their early years, agencies like EPA and OSHA reacted to hazards that were acute and obvious. In recent years, however, Congress has moved in the direction of laws that mandate preventive regulation. The Toxic Substances Control Act of 1976 provides a good illustration of a law that allows the EPA to require manufacturers to verify the safety of certain chemical products before they can be produced or sold. Consequently some chemicals are judged guilty until proven innocent, a major change from past practice. This change is significant because once a product is found to be harmful, it ordinarily takes an enormous amount of scientific evidence to justify its removal. Premarket testing should be much less expensive and time-consuming. If the toxic substance law is only the first of many similar ones, then industrial opposition to social regulation is likely to skyrocket. New and relatively inexpensive tests can be used to invalidate the use of substances and products that industry has spent millions of dollars to create.

Second, for economic regulators, which have been provided a great deal of flexibility, the discovery of undesirable situations can be handled through further study, negotiation, or a decision not to regulate at all. For social regulators, however, their flexibility is often limited because regulation is legally mandated when harmful situations are discovered. Thus because new testing methods and monitoring devices are making it much easier to discover potentially harmful products and because such products are being discovered at a quicker rate than in the past, more rather than less social regulation will occur in the future. This is in direct contrast to many economic regulations, which many policy makers now view as superfluous.

Another reason to expect greater social regulation stems from public demand. Improved levels of health, safety, and environmental protection are values that citizens applaud and that politicians cannot safely oppose. Furthermore as the public becomes more aware of its increasing exposure to carcinogens, toxic pollutants, and other hazardous products, the demands for governmental action are likely to expand.

Many opinion surveys have documented the public's probable support for increased social regulation. Surveys conducted throughout the 1970s show that a clear majority of the public favors increased governmental spending on programs to reduce air and water pollution.[25] In another survey conducted in 1977, an incredibly high 30 percent of the respondents said that no pollution of any kind should be allowed, and an additional 46 percent said that some pollution should be allowed but not so much that recreation and natural resources are interfered with or that public health is endangered. More than half the respondents in a further survey in mid-1978 agreed that "protecting the environment is so *important* that requirements and standards cannot be too high, and continuing improvements must be made *regardless* of cost."[26] In addition to their concern about the environment, most Americans believe that manufacturers who care little for consumers continually cheat them on product quality and safety. Because of this finding, the Harris Survey concluded that the country is in the midst of "a deep-seated consumer revolt, which is likely to grow and have a greater impact in the years ahead."[27] The results of these surveys suggest the public's willingness to accept a growth in social regulation. Should the number of social regulations increase, regulated groups would have still more reason to oppose them.

If social regulation is likely to increase, then must the public be wary of unchecked bureaucrats and a loss of accountability? Alternatively if social regulation is likely to increase in scope, should the Congress set standards so that elected representatives (and not appointed bureaucrats) will be making judgments about acceptable levels of health, safety, and environmental quality? Congressional choice has certain appeal, and many people have made a persuasive argument in its favor.[28] Despite the strength of their stance, however, an equally compelling case can be made that congressmen lack the time and technical expertise to become intimately involved in the implementation of social regulations. Congressional involvement could lead to an agency's inability to incorporate new scientific data, as well as its reluctance to act quickly in the absence of specific direction. These problems do not mean that Congress should abdicate its responsibility but that it could enhance its role by providing social regulators with a better indication of its intent. For example, many of the EPA's problems over sulfur oxides could easily have been avoided if the Congress had specified in the 1960s what it meant by adequate margins of safety or the language that created the controversy over nonsignificant deterioriation. Similarly the Congress could offer further assistance by detailing the relative weights that regulators should assign to social, economic, and environmental values. Congress typically indicates that it supports each of these values without favoring one over the other. A ranking of the values would facilitate a regulatory agency's task, provide explicit recognition of congressional intent, and avoid the need to debate congressional vetoes or whether an agency has exceeded its authority.

Congress could also ensure that social regulators have sufficient resources to accomplish their responsibilities. Today there is often no relation between the cost of regulation and the amount of money spent on research to solve the problems created by regulation. Some people have suggested that the total expenditures on research should equal about 3 to 5 percent of the estimated cost of regulation. Had this formula been applied to research on sulfur oxides and means of controlling them, between $1 billion and $2 billion would have to be spent over a five- to ten-year period.[29] Only a small fraction of this money has been appropriated. Adequate expenditures are likely to improve the quality of any regulatory agency's work. This is especially true for the EPA because much of the criticism directed at it stems from what people believe to be the poor quality of its research programs.

More explicit direction and better financing of social regulators are only two possible improvements, but perhaps they are the most important. Unfortunately neither is likely to occur on a widespread scale. There are too many incentives to pass vaguely worded laws, and congressional reliance on incremental budgeting makes large increases in appropriations improbable. The likely failure of Congress to remedy the problems of social regulation will not be without its consequences. As harmful substances are discovered at a more rapid pace, there will be greater public demands for more effective regulation. But if the present trends continue, both Congress and social regulators will be ill prepared to respond unless the knowledge gained from past experiences improves the initial design of new programs.

Notes

1. *New York Times,* January 23, 1977, p. 16.
2. Comptroller General of the U.S., *Federal Programs for Research on the Effects of Air Pollutants* (1975), p. 17.
3. *Congressional Record,* daily edition, April 27, 1978, pp. H3295-3297.
4. U.S. House of Representatives, Committee on Science and Technology, Subcommittee on the Environment and the Atmosphere, hearings, *Research and Development Related to Sulphates in the Atmosphere,* 94th Cong., 1st sess., 1975, p. 857.
5. Comptroller General, *Federal Programs,* pp. 33-34.
6. Ibid.
7. U.S. Senate, Committee on Public Works, Subcommittee on Environmental Pollution, Committee Print, *An Overview Inquiry of the Office of Research and Development of the Environmental Protection Agency,* 93rd Cong., 2d sess., 1974, p. 47.
8. Edwin C. Hargrove, *The Missing Link* (Washington, D.C.: The Urban Institute, 1975), pp. 112-113.
9. U.S. House of Representatives, Committee on Interstate and Foreign Commerce, Subcommittee on Oversight and Investigations, Subcommittee Print, *Federal Regulation and Regulatory Reform,* 94th Cong., 2d sess., 1976, p. 461.
10. U.S. Senate, Committee on Appropriations, hearings, *Department of Housing and Urban Development, and Certain Independent Agencies Appropriations,* 95th Cong., 1st sess., 1977, p. 593.
11. House, *Sulphates in the Atmosphere,* p. 621.
12. Senate Committee Print, *Overview Inquiry,* p. 55.
13. U.S. Congress, Office of Technology Assessment, *A Review of the U.S. Environmental Protection Agency Environmental Research Outlook, FY 1976 through 1980* (1976), pp. 23-24.
14. Comptroller General, *Federal Programs,* p. 24.

15. U.S. House of Representatives, Committee on Science and Technology, Subcommittee on the Environment and the Atmosphere, Committee Print, *Organization and Management of EPA's Office of Research and Development,* 94th Cong., 2d sess., 1976, p. 39.

16. U.S. House of Representatives, Committee on Appropriations, Report No. 92-1175, *Department of Agriculture–Environmental and Consumer Protection Appropriation Bill, 1973,* 92d Cong., 2d sess., 1972, p. 65.

17. National Research Council, *Decision Making in the Environmental Protection Agency* (Washington, D.C.: NAS, 1977), p. 13.

18. U.S. House of Representatives, Committee on Science and Technology, Subcommittee on the Environment and the Atmosphere and Committee on Interstate and Foreign Commerce, Subcommittee on Health and the Environment, joint hearings, *The Conduct of the EPA's "Community Health and Environmental Surveillance System" (CHESS) Studies,* 94th Cong., 2d sess., 1976, p. 116.

19. Ibid., p. 115.

20. National Research Council, *Research and Development in the Environmental Protection Agency* (Washington, D.C.: NAS, 1977), p. 15.

21. Murray Edelman, *The Symbolic Uses of Politics* (Urbana: University of Illinois Press, 1967), p. 24.

22. Marver H. Bernstein, *Regulating Business by Independent Commission* (Princeton: Princeton University Press, 1955), p. 90.

23. James Q. Wilson and Patricia Rachal, "Can the Government Regulate Itself?" *Public Interest* 46 (Winter 1977):4.

24. Paul H. Weaver, "Regulation, Social Policy, and Class Conflict," *Public Interest* 50 (Winter 1978):51.

25. Gladwin Hill et al., *Protecting the Environment: Progress, Prospects, and the Public View* (Washington, D.C.: Potomac Associates, 1977), pp. 26–27.

26. The results of these surveys are reported in Harry W. O'Neill, "Environmental Quality: The Public's Position" (paper presented at the Environmental Industry Conference, Washington, D.C., February 1977), and Robert Mitchell, "Information Sheet on the National Environmental Survey" (1978). Mitchell's survey was conducted by the Bureau of Social Science Research, Washington, D.C., for Resources for the Future.

27. Harris Survey, "Quality of Life," November 21, 1977.

28. See, for example, Theodore Lowi, *The End of Liberalism* (New York: W.W. Norton, 1969), and Edwin T. Haefele, *Representative Government and Environmental Management* (Baltimore: Johns Hopkins University Press, 1973).

29. House, *Sulphates in the Atmosphere,* p. 856.

Index

About the Author

Richard J. Tobin is assistant professor of political science at the State University of New York at Buffalo. He received the B.A. from the Pennsylvania State University and the M.A. and the Ph.D. from Northwestern University. Prior to assuming his present position, he was a research associate at the Center for the Study of Environmental Policy, the Pennsylvania State University. Professor Tobin has published in *Environmental Affairs*, the *American Politics Quarterly*, the *Western Political Quarterly*, and the *American Journal of Political Science*, and he has contributed chapters to several books.

Lo fingido verdadero
Acting Is Believing

Actor holding the scepter of a tragic king. Wall painting from Herculaneum,
1st century A.D., copy of 4th century B.C. Greek original,
Museo Nazionale, Naples (Erich Lessing/Magnum Photos).

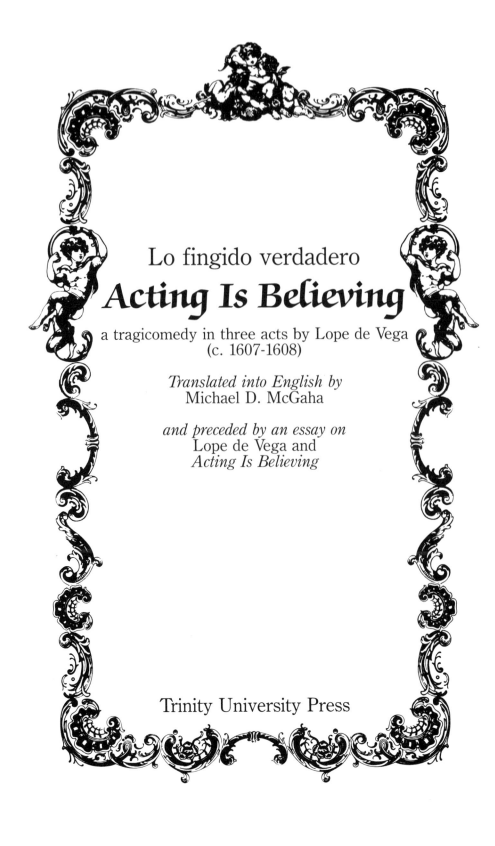

Lo fingido verdadero
Acting Is Believing
a tragicomedy in three acts by Lope de Vega
(c. 1607-1608)

Translated into English by
Michael D. McGaha

and preceded by an essay on
Lope de Vega and
Acting Is Believing

Trinity University Press

Trinity University Press gratefully acknowledges the
assistance of the *Program for Cultural Cooperation
Between Spain's Ministry of Culture and North American
Universities* in making this publication possible.

Library of Congress Cataloging-in-Publication Data

Vega, Lope de, 1562-1635.
  Acting is believing.

  Translation of: Lo fingido verdadero.
  Bibliography: p. 37
  1. Diocletian, Emperor of Rome, 245-313 – Drama.
2. Genesius, of Arles, Saint, d. ca. 303 – Drama.
I. McGaha, Michael D., 1941-   . II. Title.
PQ6459.F56   1986   862'.3   85-28947
ISBN 0-939980-14-2

Printed in the United States of America
Printed by Best Printing Company
Bound by Custom Bookbinders

Trinity University Press, 715 Stadium Drive
San Antonio, Texas 78284

Contents

Lo fingido verdadero

Acting Is Believing

Introduction: Lope de Vega and *Acting Is Believing*

I can think of no play that could better serve as an introduction to Lope de Vega's theater than *Acting Is Believing (Lo fingido verdadero)*. It is at once a manifesto of Lope's dramatic theory, an invaluable documentary record of Spanish theatrical practice in his time, an embodiment of the playwright's most deeply held views on such important topics as religion and politics, and, above all, a superb example of his dramatic craftsmanship. Written when Lope was about forty-five years old and at the height of his creative powers, *Acting Is Believing* deserves a place of honor alongside *Fuente Ovejuna, Peribáñez, Punishment without Revenge,* and *The Knight from Olmedo* as one of his most characteristic and memorable works. It is indeed arguable that *Acting Is Believing* has a greater appeal for modern audiences than those other, better known works. It is one of the ironies of literary history that Jean Rotrou's *Le Véritable Saint Genest* (1645), based principally upon *Acting Is Believing* and in many ways inferior to the Spanish original, has become recognized as one of the classics of world theater, while Lope's play has practically fallen into oblivion even in his native Spain. I hope that the present translation will serve in some measure to redress that injustice.

1

Lope de Vega's life is more fully documented than that of any other Spanish writer of his time. The earliest biography of the playwright, written by his friend and disciple Juan Pérez de Montalván, was published the year following his death (1636). Though written in a spirit of uncritical admiration and containing numerous inaccuracies and exaggerations—many of which undoubtedly originated with Lope himself—it has nevertheless remained the indispensable starting point for all later biographies. In spite of its shortcomings, it contains some priceless anecdotes which Montalván must have heard directly from Lope, as well as revealing observations on Lope's habits and idiosyncrasies by one who knew him intimately. The survival of more than 500 personal letters dating from the years 1604-1633 makes it possible to fill in many of the gaps in Montalván's biography. Never intended for publication, these letters are often brutally frank and contain passages so salacious that they were long considered unprintable. During the past century, patient research by scholars in Spanish archives has turned up many additional documents which shed light on Lope's life, the most important of these being the transcripts of his trial for libel in 1587-88 and of the hearings held in 1595, which resulted in the remission of the last two years of his sentence of exile from Madrid.

Still, the richest source of insight into Lope's life and character is his literary creations. Rarely if ever have art and life been so intimately entwined as in the case of Lope. His own experience of life provided the deepest vein of inspiration for his works, in which he constantly appeared under the most transparent of literary disguises, which fooled, and were intended to fool, no one. In dealing with these materials the scholar must of course be extremely careful—more careful than Lope himself was—to separate fact from fiction. The works which have offered the greatest yield of information to Lope's biographers are the early ballads, *La Filomena* (1621), the novel in dialogue form *La Dorotea* (1632), and the three eclogues, *To Claudio* (1632), *Amarilis* (1633) and *Filis* (written in 1635 and published posthumously). Despite this wealth of data—or perhaps because of its staggering abundance—no truly satisfactory biography of Lope de Vega has yet been written. The scholarly ones quickly become bogged down in a lifeless recitation of facts and dates, fail to separate the significant from the trivial, and read with all the excitement of a laundry list or a checkbook register. The cheap sensationalism and wild speculation of the popular biographies, on the other hand, make such icy objectivity seem bracing by contrast. Meanwhile, the enigmatic Lope de Vega continues to elude us, receding endlessly and tantalizingly behind his many masks.

Lope de Vega was born in a rented house in a working-class district of Madrid on December 2, 1562. His father, an embroiderer by trade and a native of the Santander region on the northern coast of Spain, had recently settled in the new capital city, attracted, no doubt, by the boom created by the establishment of Philip II's court there. The Madrid of Lope's formative years might perhaps best be compared to the Gold Rush towns of nineteenth-century America. It was a dirty, uncouth, brawling town filled to overflowing with adventurers, prostitutes, soldiers, courtiers, merchants, bankers, and tradesmen, all bent on getting rich quick. It was in this atmosphere that the young Lope imbibed the recklessness, insatiable ambition, and zest for life in all its sensual richness and color that were to be his trademarks.

It appears, however, that the Vega family's dreams of instant wealth or even bourgeois comfort were soon dashed. Their meager resources strained to the limit by the birth of a daughter in 1565 (Lope had an older sister, and he seems also to have had two brothers, at least one of whom was older than he), they sent the child Lope to stay for a while with his uncle the Inquisitor Miguel del Carpio in Seville. Back home in Madrid, Lope soon revealed an intellectual precociousness which won him admission as a scholarship student to a Jesuit high school, where he received the beginnings of a good classical education. Impressed by the youth's ability and industriousness, his teachers perhaps brought him to the attention of Jerónimo Manrique de Lara, Bishop of Avila, who gave him employment in his household, probably as a page, when Lope was about fourteen. The bishop too was soon captivated by the lad's intelligence and piety, and, foreseeing a distinguished ecclesiastical career for him, financed his studies at the University of Alcalá de Henares. For reasons unknown, Lope left the university and the bishop's service without completing a degree. Bursting with adolescent energy, he may have found the religious life too constraining. Possibly his return to Madrid was occasioned by the sudden death of his father in August 1578. If so, he was to be of little consolation to his widowed mother. It was probably around this time that he and his friend Hernando Muñoz, impatient to "see the world," set out on an escapade that took them as far as Astorga, some 250 miles northwest of Madrid. Soon overcome by homesickness, they began the return trip but ran out of funds by the time they reached Segovia. Attempting to sell a chain, they were arrested on suspicion of theft, but a kindly magistrate realized that their only crime was youthful imprudence and had the police return them to their families.

If we are to believe the uncorroborated testimony of *La Dorotea,* Lope then took up residence in the home of a wealthy female relative, whose hospitality he promptly repaid by getting her niece and ward pregnant. Though Lope would gladly have married the girl, her aunt found a more suitable match for her in the person of a rich, elderly lawyer. Driven from Madrid by the scandal surrounding his lover's pregnancy, Lope appears to have fled to Salamanca, where he is thought to have resumed his interrupted university studies between 1580 and 1582. Once again, however, he failed to graduate, this time leaving the university for Lisbon, where he enlisted in a naval expedition to conquer the island of Terceira in the Azores, the last stronghold of Portuguese resistance to annexation by Spain. The campaign was a short and successful one, and by fall 1583 Lope was again back in Madrid.

Now began what was to be one of the most traumatic experiences of his life. He met and fell in love with the beautiful blonde actress Elena Osorio. Elena was the daughter of Jerónimo Velázquez, manager of a theatrical troupe, and was married to the actor Cristóbal Calderón, who, however, seems to have been absent from Madrid during most of the time that Lope was involved with his wife. In any case Calderón cannot have been an overly vigilant husband, since Lope was neither the first nor the last of Elena's lovers. Lope's acquaintance with Elena coincided with the beginnings of his career as a playwright, and her father may in fact have acquiesced in the affair as a means of obtaining the young dramatist's plays for his company.

Though Elena's physical endowments were by all accounts extraordinary, they were equalled or perhaps even surpassed by her lively intelligence. She could hold her own in conversation with the wittiest wags of Madrid's literary and artistic in-crowd, and she was a remarkably talented singer, musician, and dancer. Elena's flirtations drove Lope wild with jealousy. The countless ballads in which he immortalized his love for "Filis"—his favorite poetic name for Elena—enable us to follow the vicissitudes of their intense and often stormy relationship in extraordinary detail.

The reality of Elena's love was not enough for Lope. The erotic experience somehow failed to seem fully real until it had been validated through the medium of his art. Furthermore, Lope's penchant for self-dramatization required an audience. Thanks to the wide circulation he gave the ballads, Madrid was able to follow the Lope-Elena story with the same prurient eagerness with which today's audiences tune in to a soap opera. The resulting scandal was too much for Elena's mother. Feeling that her daughter was wasting the best years of her life on the impecunious poet and that the publicity surrounding their relationship might make Elena a laughingstock and drive away more lucrative prospects, Inés Osorio determined to bring the affair to an end. Nagged and cajoled by her mother, Elena vacillated between her genuine affection for Lope and her exasperation at his jealous demands, now driving him away, now tearfully accepting him back. Lope, for his part, alternately berated Elena and humbly begged her forgiveness.

The situation took a drastic turn for the worse in the spring of 1587, when Elena became the mistress of the wealthy nobleman Francisco Perrenot de Granvela, who had recently returned to Madrid after a term as Rudolph II of Austria's ambassador to the Venetian republic. Elena seems to have done this mainly to placate her grasping mother, and it appears that Lope secretly shared her favors with Perrenot de Granvela for several months. Lope was driven to the depths of despair by his inability to persuade Elena to abandon his hated rival. Unable to bear the shame and humiliation of his position any longer, Lope at last broke with Elena. The pain of heartbreak soon gave way to rage and lust for vengeance. Lope now vented his frustration in an obsessive campaign to destroy Elena and her family. After a vain attempt to lure her father's actors away from his company, he began to circulate vicious satires on Elena and her relatives, replete with the most explicitly scabrous allusions to their character and behavior.

Fed up with Lope's outrageous impertinence, which he now saw as a danger to his livelihood, Jerónimo Velázquez filed suit for libel. Lope was arrested while attending a play performance on December 29, 1587. When called to testify, Lope flatly denied having written the satires and claimed that Velázquez's anger was due to the fact that Lope had recently been having his plays produced by Gaspar de Porres, Velázquez's competitor. Affecting the pose of a gentleman, Lope stated that he wrote these plays "as a pastime." After a thorough investigation, Lope was found guilty as charged and sentenced to four years of exile from Madrid and two years of exile from the kingdom of Castile. In a last desperate attempt to escape punishment, he forged a love letter from Elena to himself and threatened to forward it to her husband unless she withdrew the charges against him. Elena's father called his bluff and brought new charges of forgery and continued defamation of character. A search of Lope's jail cell failed to turn up the letter, which Lope claimed to have destroyed in a fit of anger. He now publicly stated, for the first time in the trial, that he had been Elena's lover for more than four years. The court then doubled his sentence to eight years of exile from Madrid and four years of exile from Castile, any infraction of the sentence to be punished by death.

Hardly had Lope been released from jail to begin serving his sentence when he committed another characteristically reckless act. When he left Madrid, he took with him Isabel de Urbina, daughter of the prominent courtier Diego de Urbina. He seems to have plunged headlong into his romance with Isabel on the rebound, shortly after breaking with Elena. Unemployed and in exile, however, Lope was in no position to support himself, let alone a mistress, and the unfortunate Isabel soon had to return home in disgrace. Her family resigned themselves to the inevitable, and Lope and Isabel were married by proxy on May 10, 1588.

Lope soon saw a way of earning some money and possibly glory as well. For the second time in his life, he made his way to Lisbon, this time to enlist in the "Invincible Armada" which set sail for the invasion of England on May 29, 1588. According to Montalván, Lope was accompanied by a brother whom he had not seen for years, but the joy of the reunion was short-lived, for this brother died in Lope's arms after having been wounded in a skirmish with some Dutch ships. After the Armada's ignominious defeat, Lope's galleon, the *San Juan*, limped its way back to La Coruña. Lope went from there to Toledo, where he was joined by Isabel and by his friends Claudio Conde and Gaspar de Porres, who accompanied the couple to Valencia, where they had decided to settle for the term of Lope's exile.

Valencia was a logical choice. Besides being a prosperous seaport, it was an important cultural center which was especially renowned for its flourishing theatrical tradition. Lope had the opportunity to get acquainted with such distinguished playwrights as Andrés Rey de Artieda, Cristóbal de Virués and Francisco Tárrega, all of whom were in their mid- to late thirties and were vigorously engaged in experiments in adapting classical and foreign dramatic forms to Spanish taste. Scholars have not yet fully succeeded in sorting out the complexities of how much Lope was influenced by the Valencian playwrights and to what extent he influenced them. It was certainly during this period that Lope began his dramatic career in earnest and first succeeded in earning a modest living from his writing. The relatively tranquil years in Valencia were at once an apprenticeship in his craft and a time of personal growth. He had now abandoned whatever hopes he may previously have cherished of a glorious military career. Looking back ruefully on the dissipation of his early manhood, he found contentment as a devoted young husband and aspiring writer.

It may have been Isabel's pregnancy that convinced Lope of the necessity of finding a more secure source of income. In 1590 he obtained employment as secretary to Antonio Álvarez de Toledo, Duke of Alba, and he and Isabel took up residence in the ducal palace in Alba de Tormes, about twenty miles southeast of Salamanca. It was there that Isabel gave birth to their first child, Antonia, named in honor of Lope's employer. Lope had hardly taken up his new duties, however, when Don Antonio was arrested and imprisoned for having jilted his fiancée Catalina Enríquez de Rivera, daughter of the powerful Duke of Alcalá, and for having married Mencía de Mendoza y Enríquez, daughter of the Duke of Infantado, without the king's consent. Until Philip II finally pardoned Don Antonio in the spring of 1593, his bastard half brother, Don Diego de

Toledo, presided over the court at Alba. Don Diego died of wounds received in a bullfight organized to celebrate his brother's return from prison.

Lope's duties as secretary cannot have been very onerous, for he found ample time to write a number of plays which were performed in Madrid during those years, many poems commemorating the duke's life and loves and, most importantly, the pastoral romance *La Arcadia*. *La Arcadia,* a *roman à clef* which treats the incidents surrounding Don Antonio's marriage and in which Lope and the other courtiers appear under pastoral pseudonyms, beautifully captures the spirit of the highly cultivated little ducal court. Light-years removed from the vulgar excitement of Madrid, Alba was a Renaissance utopia, where vast inherited wealth enabled men and women of superior intellect and refined manners to indulge their sophisticated tastes in literature, music, and art. Imitating the exquisite emotional crises suffered by the characters in their favorite reading, the pastoral romances, the courtiers whiled away the time in elaborate erotic rituals in which they posed as lovesick shepherds and shepherdesses. Nothing could have delighted and flattered them more than to have their own literature-inspired lives transposed back into the realm of art in Lope's *Arcadia*.

Harsh reality soon intruded upon Lope's participation in this escapist idyll. Isabel became gravely ill during her second pregnancy. Lope's devoted efforts to restore her to health were in vain, and she died early in 1595, after giving birth to their second daughter, Teodora. The infant survived her mother by less than a year, and their older daughter, Antonia, seems to have succumbed to a childhood disease around this same time. Unable to bear the constant reminders of his beloved wife and children, Lope now swallowed his pride and contacted Jerónimo Velázquez, begging him to intercede with the Madrid authorities to remit what remained of his term of exile. Velázquez consented and obtained the pardon.

Shortly after his return to Madrid, Lope became involved in a love affair with a wealthy young widow named Antonia Trillo, with the result that the couple were indicted for concubinage in 1596. It is not known whether the case ever came to trial. The scandal surrounding the indictment may have caused the Duke of Alba to dismiss Lope from his employment. Although his plays were by now in considerable demand, Lope's hopes of supporting himself as a playwright were dashed when Philip II ordered the theaters closed in mourning for the death of his daughter Catalina. Lope was forced to find a new secretarial position, this time in the household of Francisco de

Ribera Barroso, Marquis of Malpica. Lacking a market for his plays, he began work on *La Dragontea,* an epic poem based on the life and exploits of Sir Francis Drake, which was published in 1598. For reasons unknown, Lope had by then again changed jobs and was working as secretary to Pedro Fernández de Castro, Marquis of Sarria.

The year 1598 was to be an eventful one for Lope. On May 3 he was married to Juana de Guardo, daughter of Antonio de Guardo, a rich meat and fish merchant. Juana brought Lope a handsome dowry, and his detractors were quick to accuse him of having married her for her money, a charge which most of Lope's biographers have seconded. In the same year Lope published his *Arcadia,* prefacing it with a spurious coat of arms in which he claimed descent from the legendary medieval hero Bernardo del Carpio. This piece of childish vanity attracted the ridicule of several literary notables of the day and particularly of the poet Luis de Góngora, who was to carry on a long and bitter feud with Lope.

Even before his marriage to Juana de Guardo, Lope had embarked on a passionate affair with the famed actress Micaela de Luján. Micaela was married to the actor Diego Díaz, but he seems to have gone to seek his fortune in Peru before Lope met his wife. His timely death in Peru a few years afterward surely spared him much embarrassment. Lope's marriage apparently interrupted his liaison with Micaela only very briefly, if at all. For the next ten years he would maintain two households, dividing his time between his wife and his mistress.

A blonde like Elena Osorio, Micaela was extraordinarily beautiful. Though she was in fact illiterate, Lope tried to pass her off as a talented poet, writing poems — often in praise of himself — signing them with her name, and publishing them in several of his books.

In the spring of 1599 Lope accompanied his employer, the Marquis of Sarria, to Valencia to attend the double wedding of the new king, Philip III, to the Archduchess Margarita of Austria, and of the king's sister, Isabel Clara Eugenia, to the Archduke Albert. The lavish festivities, which coincided with the annual Carnival, went on for weeks, and Lope had a prominent part in them. We can well imagine his delight when the king, on the day before his wedding, lifted the ban on the performance of plays which had been in effect for a year and a half. On the wedding day itself, his one-act allegory *The Wedding of the Soul and Divine Love* was performed.

Around the time of Lope's return to Madrid, his wife gave birth to their daughter Jacinta, who was christened on July 26, 1599. Nothing more is known of this daughter, who probably died in early childhood. Perhaps now assured of being able to earn a living from his plays, Lope resigned his position with the Marquis of Sarria in 1600. The removal of the court to Valladolid that year, however, left Madrid desolate and depressed. Hence, when his lover Micaela went to Seville in 1602, probably under contract to perform there during the spring theater season, it is not surprising that Lope decided to join her. He soon formed close friendships with some of the city's leading intellectuals, frequently attending the literary soirées held in the elegant home of the aristocratic poet Juan de Arguijo.

Lope was back in Madrid by the autumn of 1602, when he published two more books, the long poem *The Beauty of Angelica* and his *Rhymes*. He divided most of the year 1603 between Madrid and Toledo, but had rejoined Micaela in Seville by October 19 of that year, the date on which their son Félix was baptized. Micaela was now about thirty-five years old, a widow, and Félix was her seventh child. Lope probably fathered all of these but the two eldest daughters, Agustina and Dionisia. Lope remained in Seville through the spring of 1604. His Byzantine romance *The Pilgrim in His Homeland* was published in Seville that year. As part of the preliminaries to the novel, he published a list of 219 plays he had written by that date.

In the summer of 1604 Lope and Micaela moved to Toledo. His decision may have been motivated by false rumors that the court was soon to locate there. As soon as he had settled Micaela and the children in a rented house, he dashed off to Madrid, where he managed to talk Juana, now in an advanced state of pregnancy, into moving to Toledo. With his two households within a few blocks of each other, he was now able to simplify the logistics of his domestic life. Poor Juana's second child seems to have met the same fate as her first. Micaela, however, bore Lope another daughter, Marcela, in May 1605. That same month the Toledo City Council honored Lope by inviting him to preside over a poetry contest held in celebration of the birth of the future Philip IV. Besides continuing to turn out plays at an astonishing rate, Lope was now at work on his most ambitious epic poem, *Jerusalem Conquered.*

During a visit to Madrid in August 1605, he met the twenty-three-year-old Luis Fernández de Córdoba y Aragón, Duke of Sessa, who was to be his most constant patron for the rest of Lope's life. Lope continued to reside in Toledo throughout the year 1606, and Juana gave birth to their son Carlos Félix there that spring. The court had now returned to Madrid, and Micaela seems to have moved there sometime in 1606. On January 28, 1607, she bore Lope a son, whom she named after his father, and Lope attended the christening in Madrid on February 7. Nothing more is known of Micaela after the birth of little Lope. It is assumed that she must have died soon thereafter.

By 1609 Lope, now forty-six years old, had won great fame and considerable wealth as a playwright. The Duke of Sessa had taken him on as his secretary and confidant, and this connection gave Lope a certain entrée into Madrid's aristocratic society. These enviable attainments, however, were unable to satisfy Lope's longing for honor and recognition. In spite of the immense popular success of his plays, intellectual critics continued to scorn them as lacking in taste and elegance, fit only for the vulgar rabble. They were particularly incensed by Lope's failure to follow the dramatic precepts laid down by Renaissance commentators on Aristotle's *Poetics.* Though on one level Lope was fully aware of the magnitude of his achievement as a dramatist, he was at the same time a deeply insecure person with a fragile ego and hence was very susceptible to these criticisms. These mixed emotions explain why, when he penned a reply to his critics in his *New Art of Writing Plays,* published in 1609, he posed on the one hand as an intellectual who admired the classical dramatic tradition but was driven by economic necessity to write flawed plays which catered to the popular taste, but on the other hand defended his innovations, such as the mixture of comedy and tragedy, as rooted in the Aristotelian principle of verisimilitude. Conscious of the onset of age, Lope was profoundly troubled by the place he would eventually occupy in literary history. Feeling that his plays were too vulnerable a claim to fame, he pinned his hopes for immortality on the epic *Jerusalem Conquered,* also published in 1609. Written with great attention to style and full of erudite allusions, the *Jerusalem* has ironically long ago passed into oblivion, while Lope's best plays are as vibrant and engrossing today as they were almost four centuries ago.

As Lope approached the age of fifty, he was increasingly preoccupied with a yearning for respectability and perhaps also gnawed by feelings of guilt for his past immorality. In 1609 and 1610 he joined two religious confraternities in Madrid. In 1610 he moved to Madrid, purchasing a handsome two-story townhouse with a charming garden which was to offer him much pleasure and relaxation during the remainder of his life. Comfortably settled in his new home with Juana and little Carlos, he looked forward to enjoying the fruits of his labors in a serenity he had not known since the early years in Alba with Isabel.

Once again, however, his dreams of contentment were cruelly shattered. Several unrelated events conspired to destroy his newfound peace and prosperity. In June 1611 his patron the Duke of Sessa was exiled from Castile as punishment for having engaged in an altercation with the police during a night of carousal. Juana was seriously ill throughout the summer, and Lope spent many anxious hours at her bedside. Though his marriage to Juana was no love match, he does seem over the years to have developed a sincere affection and admiration for her. Carlos fell ill with a fever in September but had recovered by the end of the month. Lope's remorse of conscience, perhaps aggravated by the illnesses of his wife and son, tormented him, and he poured out his sorrow and longing for divine forgiveness in a volume of *A Soul's Loving Soliloquies to Its God*. He also joined the Third Order of St. Francis, an association of devout laymen, on September 26.

The following month the government ordered the theaters closed in mourning for the death of Queen Margarita. The queen's demise was especially painful to Lope, for she had been an ardent theater buff and had often sponsored performances of his plays at court. In November Juana's condition took a turn for the worse. Around this time someone attacked Lope with a knife. Lope managed to knock his assailant down and escaped with nothing worse than a cape slashed to shreds and a bad case of nerves. Towards the end of February 1612, Lope fell and badly injured his arm, and Juana suffered a miscarriage. That summer six-year-old Carlos, whom Lope adored, died after a brief illness. Juana became pregnant a few months later and on August 1, 1613, gave birth to a daughter. Nine days later, Juana died.

Hoping to find distraction from his misery, Lope left Madrid in mid-September to accompany the royal family on a trip to Segovia, Burgos, and Lerma. The all-powerful Duke of Lerma, Philip III's favorite, entertained the court with a lavish production of one of Lope's plays—probably his *Perseus*—in the gardens of his country estate at Ventosilla. Lope took every opportunity during this trip to persuade the duke to appoint him to the prestigious post of court chronicler, but his ambitions were once again thwarted. With little interest in literature and even less in theater, the king, who was a bit of a prude, seems to have disapproved of Lope.

Back in Madrid, Lope sank into a deep depression bordering on despair. Looking back over the half-century he had lived, he saw only disappointment, desolation, and shame. His once bright hopes seemed to have ended in utter failure. The intelligentsia, whom he so yearned to impress, looked upon his work with disdain. Even the masses, who had lionized him for so many years, now began to find his plays old-fashioned and sought out the novelties offered by a younger generation of playwrights. The aristocracy, offended by his impatient eagerness to gain entrance to their select inner circles, scorned him as a vulgar parvenu. His finances were in a state nearing bankruptcy, reducing him to an ever more humiliating dependence on the Duke of Sessa, whose favors were dearly bought in flattery and coaxing. No longer could he find stoic consolation in the awareness of his own superiority, for his conscience was haunted by reminders of his past misdeeds. The deaths of Micaela, Juana, and Carlos inevitably brought him face to face with his own mortality. Convinced that the calamities that had befallen him were divine retribution for his sinful life, Lope now felt an urgent need to make his peace with God.

As a youth, he had studied for the priesthood at Alcalá de Henares, and he seems also to have gone to Salamanca a few years later with a renewed intent to pursue a religious vocation. It now seemed clear to him that all of his misfortunes had resulted from his failure to remain steadfast in that vocation. Determined to devote the remainder of his life to making reparation for his wasted youth, Lope received minor orders in Madrid in March 1614, then moved to Toledo to prepare himself for ordination to the priesthood. It was no easy task to convince church officials of the depth and sincerity of his change of heart. Impatient as usual, he was dismayed by the delays and obstacles they placed in his path. Nevertheless, he kept busy by studying for his examination, writing devotional poetry, and shopping for decorations for his private chapel. Finally he was ordained a priest on May 24, returning to Madrid early in June.

Lope hoped that an appointment as private chaplain to the Duke of Sessa would relieve him of some of the more distasteful duties he had been performing as the duke's secretary, particularly that of composing love letters for his patron to send to his many paramours. He seems even to have cherished some hopes of reforming the wayward duke. Sessa, however, would have none of it. Viewing Lope's newfound scruples of conscience as just another of the volatile poet's whims, the duke bullied him into continuing to serve as his procurer. Lope suffered much anguish over this dilemma, but in the end, unable to renounce the duke's protection and material assistance — Lope was now supporting his children by Micaela as well as his legitimate daughter Feliciana — he backed down. His continuing close relationship with the amoral Sessa, together with his ongoing involvement in Madrid's theatrical demimonde, were almost inevitably to prove fatal to his resolve to live a pious Christian life.

Lope did not succumb without a struggle. In June 1615 we find him back in Toledo, where he had fled to escape the unwelcome and embarrassing attentions of a former lover. In an outraged letter to the Duke of Sessa, he swore that he had remained chaste since his ordination. He also journeyed that summer to Avila, where his eloquent preaching won him appointment to a chaplaincy endowed by Bishop Manrique de Lara, whose patronage he had enjoyed as a youth. In October he accompanied the court to Burgos, where the double marriage by proxy of the future Philip IV to Princess Elisabeth of France, and of Philip's sister Ana Mauricia to Louis XIII of France, was celebrated. Lope and Sessa traveled in the retinue of the Spanish princess as far as the Bidasoa River, where they met Princess Elisabeth, returned to Burgos on November 22 and traveled thence to Lerma. Lope returned to Madrid in December.

It seems to have been around this time that he plunged into a tempestuous love affair with the actress Lucía de Salcedo, to whom he refers in his correspondence as "la Loca." Anxious to safeguard his reputation, Lope tried for as long as possible to conceal this affair even from the prying eyes of the Duke of Sessa. Forced at last to confide in his master, who took a voyeuristic delight in hearing every detail of Lope's amorous entanglements, he nevertheless remained uncharacteristically discreet in his references to Lucía. Lucía was probably the former lover from whose scandalous attentions Lope had fled to Toledo the previous summer. In any case she seems to have taken the initiative in the affair, shamelessly seducing the aging poet-priest. Perhaps she was amused by the challenge of overcoming his resistance. At any rate, once he was sufficiently consumed by passion for her to take the tremendous step of breaking his religious vows, she appears to have lost interest and to have taken another lover. Or maybe it was merely a case of "out of sight, out of mind." Lucía had gone on tour to Barcelona with the acting company of Hernán Sánchez. The Count of Lemos, on his way home from serving as Viceroy of Naples, took the company aboard his ship in Barcelona, and they provided entertainment for him and his retinue on the last leg of the journey to Valencia. Lope, having heard rumors of Lucía's unfaithfulness and receiving no reply to his letters to her, made a hasty and ill-conceived trip to Valencia in the summer of 1616 to have it out with her. He had hardly arrived in Valencia when, physically exhausted by the hardships of the trip in the heat of the summer and suffering emotionally from the torments of jealousy, he fell dangerously ill with a fever which caused him to fear for his life. Lucía seems to have quieted his jealous fears, however, and the pair were happily reconciled. Though still in a weakened condition, he returned to Madrid in September.

The autumn of 1616 was one of the most miserable periods of Lope's life. The flighty Lucía, aptly nicknamed "la Loca," seems to have delighted in torturing him. At one moment she would shower him with affection, convincing him of her undying love, but then at the first opportunity she would indulge in brazen flirtations with other men. She does seem to have cared deeply about Lope, or at least to have been badly in need of his love, but she was incapable of carrying on a sustained relationship with one man. Lope's impossible infatuation with Lucía came to a sudden end when, while attending a literary soirée in the garden of one of his friends in Madrid, he met and instantly fell in love with Marta de Nevares Santoyo.

Marta, a brunette with fair skin and dazzlingly beautiful green eyes, was twenty-six years old. Her parents had married her at the tender young age of thirteen to the prosperous businessman Roque Hernández de Ayala. Hernández seems not only to have been a coarse and ignorant man, but a wife-beater as well. The refined and sensitive Marta, who loved good music, poetry, and intelligent conversation, can have had little in common with him. She had apparently already sued him for a legal separation on the grounds of cruelty by the time she met Lope, but the suit seems to have bogged down in legal technicalities. Perhaps the mere act of filing the suit had been sufficiently threatening to persuade Roque to agree to a truce. At any rate, at the time of her first encounter with Lope she evidently enjoyed considerable freedom to live her life as she pleased.

As the year 1616 drew to a close, Lope wrote to Sessa: "I am as lost as ever in my life I have been for the soul and body of a woman, and God knows how I regret it, for I don't know how this can go on, nor how I can go on living without enjoying it." Lope seems to have fallen more deeply in love with Marta than with any other woman since Elena Osorio. In his great novel *La Dorotea* he would blend the traits of the two women into a single character. His irresistible attraction to Marta was due no doubt to the fact that she was the first woman he had known in thirty years who combined great beauty and personal charm with an intellectual acuteness which rivaled his own. If Lope was immediately smitten with love at first sight, however, such was not the case with Marta. She was in no hurry to sacrifice her freedom and respectability for the love of a

fifty-four-year-old priest, even if he was Spain's most famous playwright. Years later in the eclogue *Amarilis* — Lope's poetic pseudo-
nym for Marta — he recalled that

> It would be easier by far to recount
> the various flowers of this pleasant grove
> than the longings, fears and sighs
> of the hope of my sweet pain
> until, after long delays,
> I won her heart, though not her embraces.

At last, after an arduous courtship including much lavish wining and
dining — financed by the duke, who was delighted to see Lope his old
self again — Marta surrendered. As a condition of his support, the
duke insisted on being given copies of the love letters exchanged by
the couple for his private enjoyment.

By the beginning of 1617 Marta was pregnant with Lope's child.
After six years of almost unrelieved melancholy, Lope was deliri-
ously happy. His joy was somewhat marred, however, by the troublesome presence of Marta's husband, by intermittent bouts of
remorse, and by the rage of Lucía de Salcedo. Furious at Lope's
rejection of her, Lucía did everything in her power to defame him
and Marta and to make their lives miserable. When Marta had a
long and difficult labor before giving birth to their daughter Antonia
Clara on August 12, Lope irrationally blamed Lucía's malevolence.
He even suspected her of resorting to witchcraft to get revenge on
him. Roque and his family, fully aware of all that was going on, further exploited the situation by abducting the child — legally Roque's
daughter — and demanding a rich ransom for her return. This sordid
predicament ended only with Roque's death in late 1618 or early
1619. Lope's merciless jubilation over Roque's death is deplorable,
though understandable in the circumstances.

Lope and Marta now settled down to a tranquil family life with
Marcela, Lope Félix, Feliciana, and little Antonia Clara. These were
productive and prosperous years, which Lope spent happily among
his children, his books, and the flowers he lovingly tended in his garden. When the post of court chronicler became vacant in 1620, he
once again applied for the job and was again rejected, probably
because of his scandalous private life.

In February 1622 the sixteen-year-old Marcela took her vows as a
nun in the Trinitarian convent within two blocks of her father's house.
She would live there to the age of eighty-three, composing poems
worthy of her father's pen and distinguishing herself as an adminis-

trator. Though saddened by his daughter's departure from the household, Lope found consolation in the knowledge that she had escaped "the world's madness and deceits," as he put it in a poem written on the occasion of her profession. He would often have reason to envy her serenity during the troubled years that lay ahead of him.

Around the time that Marcela entered the convent, Marta suddenly lost her eyesight. Lope sought the best medical treatment then available, even procuring the assistance of an Englishwoman who was noted for her treatment of eye diseases. Although Marta did show some improvement – by April 1628 she was able to distinguish light from darkness – she never altogether recovered her sight. The fact that her blindness was followed about four years later by a long period of insanity, and then by her death in April 1632 at the early age of forty-two, has led some of Lope's biographers to surmise that she was suffering from syphilis. Perhaps Lope had contracted the disease from Lucía de Salcedo. If so, his blaming Lucía for Marta's insanity and blindness would be more than a matter of mere superstition.

Marta's illness once again aroused Lope's religious scruples. During these difficult years, in addition to trying to comfort Marta as much as possible in her afflictions, Lope worshipped daily at the church of Our Lady of Atocha and frequently visited the sick in Madrid's hospitals. He also wrote more religious poetry, publishing a volume of *Divine Triumphs* in 1625 and a religious epic poem, *The Tragic Crown,* based on the life of Mary Queen of Scots, in 1627. This last poem was dedicated to Pope Urban VIII, who generously responded by sending Lope a personal letter of thanks, granting him the title of Doctor of Theology and making him a knight of the Order of St. John of Jerusalem.

Lope had made a remarkable comeback as a playwright during the early years of his union with Marta, but by 1628 the fickle public seems once again to have tired of his work. A serious illness in the spring of that year left him worn out, discouraged, and longing to retire from the theater. With this in mind, he begged the duke to appoint him officially as his chaplain and assign him a regular salary, but Sessa enjoyed the flattery that accompanied Lope's constant requests for money too much to grant such a petition. It is no wonder that Lope was weary of writing plays. In the prologue to the twentieth volume, or *Parte,* of his dramatic works, published in January 1625, he claimed to have authored 1,070 plays. This was surely one of his customary exaggerations, but his literary output was nonetheless sufficiently staggering to justify the epithet "Monster of Nature," which Cervantes had conferred upon him in 1615.

Lope's son and namesake Lope Félix followed all too closely in his father's footsteps, causing the aged writer much embarrassment and dismay. The tenderhearted Marta had to use all of her ingenuity to keep the peace between father and son. Lope was much relieved when Lope Félix enlisted in the army and went off to fight the Dutch and the Turks, eventually rising to the rank of captain.

If Lope's popularity with the public had declined somewhat, he seems during these years to have enjoyed greater favor at the court of Philip IV. His extravaganza *The Forest without Love* was staged at the royal palace in 1629, complete with elaborate stage machinery designed by the noted Florentine engineer Cosimo Lotti; two years later, he was commissioned to write his *St. John's Night* for a gala performance before the royal family in the gardens of the Count of Monterrey. In 1630 his long poem in praise of contemporary poets, the *Laurel of Apollo,* was published. Whatever pleasure these accomplishments may have brought him, however, was soon shattered by Marta's death, a blow from which Lope never recovered.

At the end of 1632 Lope published his superb novel *La Dorotea,* in which he immortalized the two great loves of his life, Elena Osorio and Marta de Nevares. *La Dorotea* reveals an astonishing depth of self-knowledge and is of the greatest interest both as a work of art and as a psychological document. In the following year he poured out his love and grief for Marta in the eclogue *Amarilis.* On December 18, 1633, his daughter Feliciana married Luis de Usátegui, a bureaucrat employed in the offices of the Royal Council of the Indies.

Lope suffered still another tragic shock in 1634 when he received the news that his son had perished in a shipwreck while on a pearl-fishing expedition to the Antilles. His one consolation in so much sorrow was his dear daughter Antonia Clara, whose beauty and lively intelligence were constant reminders of her mother. Now blossoming into womanhood, Antonia kept house for her father, served as his secretary, and relieved the loneliness of his old age. It cannot have been a very satisfying life for a vivacious adolescent.

While attending a party with her father in late August or early September 1634, Antonia was called upon to sing. Her beauty and charm attracted the attention of a high-ranking young nobleman named Cristóbal Tenorio. The rich and handsome young man quickly swept Antonia off her feet, bribed Lope's housemaid, and won entrance to Antonia's bedroom during Lope's absences from home.. Though Lope noticed suspicious changes in Antonia's behavior—particularly an unusual attention to her appearance for no evident reason—she easily managed to quiet her indulgent father's fears. Returning home one evening, Lope called his daughter but received no reply. He then called for the maid, but again with no result. Even the dog was gone. A frantic search of the house revealed that Antonia's room was in a chaotic state, hastily stripped of clothes, cosmetics, and jewelry. Antonia had run away with Tenorio who, soon tiring of her, abandoned her to her fate. She would later return to her father's house, where she died at the age of forty-seven in 1664.

With nothing left to live for, Lope now longed insistently for the release that only death could bring. Montalván recorded the events of his last days in great detail. On August 25, 1635, he said Mass in his chapel, watered his garden, and retired to his study. Around midday he felt a chill. Nevertheless, he kept an appointment to attend a philosophical and medical discussion at the Scots' Seminary, where he fainted and had to be carried home. The king's private physician, Dr. Juan de Negrete, examined Lope and advised him that the end was near. After receiving the last rites, Lope blessed his daughter Feliciana and bade farewell to his friends. He told Montalván that he would gladly have traded all the applause he had ever received for one more virtuous deed. On August 26 he felt well enough to write two poems, a sonnet on the death of a Portuguese gentleman and a 246-line moral meditation on the theme of the Golden Age. After describing the peace and harmony which prevailed during that mythical first age of the world, he launched into a bitter jeremiad against the greed, cruelty, and lust which destroyed that earthly paradise:

> They stole the generous maidens
> for their lovers, claiming they'd be wives,
> betraying their friends,
> and all of this went unpunished:
> for many who feared
> losing the judge's staff, bent it instead;
> and others who took it up
> intending to straighten it, broke it.
>
> Oh, favor of kings!
> The stars receive their light from the sun.
> They call the laws mere cobwebs,
> which entangle the little creatures
> while the strong ones break them.
> Woe to the lord who abandons his vassals
> to cry out their just complaints to heaven!

Hence holy divine Truth,
seeing the earth in such a state —
the rich man worshipped,
the poor man wretched,
whom even death fails to favor,

for the more he cries out, the less he is heard:
the wise man despised,
the flatterer heeded and rewarded,
money triumphant,
Joseph sold by his own brother,
shame and mockery of the human condition;
and amidst this clamorous confusion
Democritus laughing,
Heraclitus weeping,
death not feared
and man building
for the dream of so short a life,
heedless of the inevitable summons —
with swift flight
mounted up on her own shoulders to heaven.

Such a world could clearly no longer hold any allure for Lope. On August 27 he awoke too weak to talk, though he remained conscious until the end. Surrounded by his friends, he died at 5:15 that afternoon at the age of seventy-two. Practically the entire population of Madrid turned out to honor him at his funeral the next day. His spirit, so restless during most of his life, had at last found repose.

2

This, then, was the man who around the year 1608 sat down in his study to write the play *Acting Is Believing.* Ostensibly a religious drama based on the life and martyrdom of St. Genesius, patron saint of actors, the play is in fact a sustained meditation on the phenomenon of role-playing and its consequences in human life. According to legend, St. Genesius was a Roman actor who, while performing a farce in mimicry of Christian ceremonies before the Emperor Diocletian, suddenly converted to Christianity and began preaching to the emperor, whereupon he was executed forthwith. Lope must at once have sensed the rich dramatic potential in the theme of the actor who identifies totally with his role. Rambunctious youth, brilliant but lackadaisical student, gallant soldier, distinguished and prolific author, passionate lover, devoted husband, loving father, pious but guilt-ridden priest—Lope had played, or was yet to play, an astonishing variety of seemingly contradictory roles in his own life, and each of these life experiences would provide grist for the mill which ground out a constant stream—a tidal wave might be a more appropriate metaphor—of poems, plays, and novels.

Lope's life can be seen as a constant attempt to substitute a pleasing and artful fantasy for a more prosaic, complex, and sometimes sordid reality. Just as it is difficult for us to understand why he felt compelled to exaggerate his accomplishments as a writer, which in unvarnished truth were so enormous as to defy belief, so the mind boggles at his need to add romance, glamour, and adventure to a life already more exciting and eventful than most fiction. Yet reality somehow always seemed too small, too confining, too imperfect, and at the same time too improbable and even *unreal* to satisfy him. Tormented by deep-rooted doubts of his own worth—which no amount of acceptance, love, or success could put to rest—he was continually driven to rewrite the script of his life, donning one mask after another in search of the one that would at last win unconditional acclaim. Though he struggled valiantly to observe decorum and to play each new role with the utmost consistency, unruly aspects of his own personality kept betraying the man behind the mask, spoiling the game of "let's pretend." Fate too kept rudely stepping in with unpleasant surprises to remind him that he was only the leading man, not the author, of this play. Though he sometimes succeeded in fooling others, he was unable to deceive himself. Deep inside, a nagging voice that would not be silenced relentlessly repeated: "This is not you; this is only play-acting," until at last Lope found the courage to accept himself in all his disorderly human complexity, warts and all, renouncing nothing. In *Acting Is Believing* Genesius tells the Emperor Carinus of an actress named Lisarda who became a Christian, to which the emperor significantly replies: "She gave up acting." Christianity implies a total honesty which precludes role-playing. It was this honesty which at last enabled Lope to write *La Dorotea,* which—despite its artistic reordering of such circumstantial details as names and dates—is as uncompromising a confrontation with self, as scathing a self-portrait as is to be found in all of literature. But if that self-portrait is brutal in its honesty, it is also wonderfully kind. For, paradoxically, it was only when he was able to acknowledge his own frailty and baseness and even to laugh good-naturedly at the arrogant poses he had affected in his youth that Lope at last succeeded in loving himself. To understand all is to forgive all.

In 1608 Lope still had far to go before attaining that level of self-awareness. Nevertheless, *Acting Is Believing* demonstrates that he had already given deep thought to the problem of role-playing. In one sense, of course, role-playing, like game-playing, is an integral part of all human societies. Each different stage in life, each social class and profession, and many different types of relationships are accompanied by the expectation of particular types of language, dress, and behavior. However, if these conventions are indeed to some extent necessary for the orderly functioning of society, they also exact a terrible cost. A heightened preoccupation with social justice and the rights of the individual has in modern times made us acutely aware of the damage done by stereotyping – the refusal to acknowledge that no individual is fully defined by his or her role. The positive results of recent changes in sex roles achieved by pressures from the women's liberation movement have brought us to a new realization of the enormous waste of human potential caused by sexual stereotyping. Watergate has reminded us of how a false sense of impunity and invulnerability can result when an individual identifies too completely with an office of power and authority. Modern psychology has shown us how growth and development can be stifled and grave psychic disturbances can occur when a person subordinates too many of his needs and desires to the narrow, procrustean demands of a single role. Modern artists and writers have rebelled against the tyranny of imitation, demonstrating that art and reality are two very different things and that too great a subjection to verisimilitude can destroy creativity. Representational painting and "realistic" theater, like role-playing, are in a sense nothing but clever frauds, attempts to pass off an illusion as reality.

The trouble with role-playing is that it always contains an element of dishonesty. The more attractive the role, the greater is the temptation to treat it as all-encompassing and ultimately real. No one ever is or should be nothing but an enterprising executive, an efficient secretary, a loving parent, an orthodox Jew, or a conservative Republican, to cite only a few examples. We are all larger, more complicated, and vastly richer in possibilities than the labels that we or others attach to ourselves.

Though we tend to think of these concerns as peculiarly modern, none of them would have come as a surprise to Lope de Vega. He knew very well that comedy and tragedy as discrete genres are too limited to serve as a paradigm of the human condition. Variety and freedom were his artistic watchwords. As a man who spent his entire adult life writing for the theater, Lope was well aware that all roles are provisional and that no single role can exhaust the potential of a given actor.

In *Acting Is Believing* he shows us two Roman emperors – Aurelius and Carinus – who foolishly took the rhetoric of power at face value, failing to understand that it was addressed not to them personally but to the office they held. The haughty Aurelius dared to consider himself the equal of Jupiter and was struck by lightning in punishment for his blasphemy. Carinus, forgetting that power was entrusted to him by his people on condition that he use it for the common good, perpetrated intolerable abuses on his subjects and was assassinated. Only then did he realize that "the law is the same for all who are born." Aper, Numerianus's father-in-law, wrongly believed that if he murdered Numerianus, his "great power, reputation, and noble birth" would ensure his own election as emperor. Diocletian, soldier and son of a slave, refused to be defined by his humble role, aspiring instead to that of emperor. Once having attained that ambition, he played his new role better than those who had been born to it, always remembering that "though my status has changed, my soul has not, nor do I think it ever will."

The actor-playwright Genesius, incapable of distinguishing role-playing from reality, writes a play which will enable him to show himself to full advantage and to cast aspersions upon Octavius, his rival for the love of the actress Marcella. The script called for Marcella to run away with Octavius. Octavius' servant Pinabellus, betraying his master, would attempt to take Marcella for himself, but Genesius would arrive in time to save Marcella, and true love would triumph. Genesius's plan backfired, however, when, while performing the play before Diocletian and his court, Marcella took advantage of the opportunity to run away with Octavius in reality. When Genesius desperately implored Diocletian to send troops to bring the lovers back, the emperor, suspecting that it was all a trick to get him to participate in the play, angrily declared the performance over. As Genesius learned to his sorrow, the roles we play and those in which we try to cast others are often futile attempts to control the uncontrollable, to predict the unpredictable.

The emperor now asked Genesius to put on a play about a Christian. Genesius began to rehearse, tentatively ad-libbing as he struggled to "get inside" the character he was to play, to experience how it feels to be a Christian. As he fumbled for words, the divine playwright began to supply them, and Genesius was perplexed to find himself saying all sorts of things that would never have occurred to him under normal circumstances. Recovering his composure, he thought that perhaps he had merely gotten carried away by his role, as had so often happened before. When the actual performance began, it was the turn of the other actors to be confused, as Genesius again and again departed from the script. By the time the actor who was to play an angel appeared, a real angel had already usurped his part. As the actors began to quarrel among themselves, Diocletian was enraged by their incompetence. When Genesius insisted that he had indeed become a Christian and that what the audience was witnessing was not feigned but real, Diocletian, reluctantly forced to play the role of emperor, sentenced him to death. Ironically, Genesius, who before had sought to make a play of his own life, now found that a role he had intended as fiction had turned into the ultimate reality.

As the other actors were interrogated, they insisted that the roles they customarily played had nothing to do with reality. Salustius, assigned to play villains, argued that he was in fact of a good family and had never betrayed anyone. When the judge jokingly asked Fabritius, who played serious roles such as fathers and kings, to trade places with him, Fabritius protested that he didn't know the laws. Only for Genesius were role and reality the same. He had learned the lesson so well enunciated by the character Howard Campbell in Kurt Vonnegut's novel *Mother Night:* "You are what you pretend to be, so be very careful what you pretend to be."

If we disregard its religious overtones—which is not hard to do, since they seem to have been introduced almost as a casual afterthought—*Acting Is Believing* speaks to us today with an astonishing freshness and relevance. Some of Lope's other masterpieces, such as *Peribáñez* and *Punishment without Revenge,* may be so deeply rooted in the culture of his time and place that contemporary audiences cannot fully appreciate them without some specialized knowledge of that culture, but *Acting Is Believing* requires no such introduction. Its theme is universal, and its treatment of that theme arouses a shock of recognition that is the hallmark of dramatic genius.

3

Acting Is Believing was first published in the *Decima sexta Parte de las Comedias de Lope de Vega Carpio* (Madrid, 1621). The title appears in the list of Lope's plays published in the sixth edition of his novel *The Pilgrim in His Homeland* in 1618, so we know that the play was already in existence by that date. The song addressed to Lucinda—Lope's poetic name for Micaela de Luján—which precedes the performance of the play-within-a-play in Act II, suggests that *Acting Is Believing* was probably written by 1608, since Micaela seems to have died sometime in 1607, and Lope ceased referring to her in his work shortly thereafter. The play's unusual preoccupation with dramatic theory offers a further indication that it was composed around 1607-1608, when Lope was at work on his *New Art of Writing Plays.*

It is clear that the character Genesius is to some extent a persona of the author. Lope ingeniously departs from tradition by making Genesius a playwright as well as an actor. Furthermore, he casts Genesius in a role that he himself coveted but never obtained – that of an official court playwright who enjoys a close and rewarding relationship with two emperors. There is a special poignance in Genesius's words to Diocletian in Act II: "When Alexander the Great arrived in Athens, he was eager to meet the great poet Thebanus, *for in those days princes honored poets.*"

Some of the play's critical pronouncements parallel and even paraphrase passages in the *New Art.* What is especially interesting is that the most conclusive arguments in defense of Lope's new dramatic formula are placed not in the mouth of his counterpart Genesius but rather in the more prestigious ones of the emperors Carinus and Diocletian. When Carinus commissions Genesius to have a play written for performance before his court, he recommends that Genesius entrust the task to a certain Aristeles. Genesius remarks that "He'll observe the rules," but the emperor objects: "No, do it just like always. I don't like to be limited by art and the precepts." Genesius cautions him that the intellectuals will be offended. Carinus insouciantly replies: "Well, let them be. Delight the ears, and that's enough, as long as there's no absurdity that can be seen." Genesius has a similar conversation with Diocletian in Act II. He suggests that Diocletian might like to see a performance of Plautus's *Miles Gloriosus.* The emperor anachronistically comments: "I'd like a story that's more original, even if less artistic. As you can see, I share the Spanish taste in plays. As long as it's believable, I'm not too picky about whether it follows the rules. In fact I find the precepts too limiting, and it's been my experience that those who are too careful to observe artistic dictates never succeed in making a natural plot." These two discussions recall Lope's admission in the *New Art:*

> Except for six, all my other plays
> have gravely sinned against art.
> Yet I stand by what I've written, and I know
> that, though they might have been better if written

differently,
they wouldn't have given the pleasure that they
    gave,
because sometimes what goes against the right
for that very reason gives delight.

Another source of irritation to Lope was the court's fondness for
silly plays whose only interest lay in their use of machinery to
achieve spectacular special effects. In the *New Art* he had blamed
his failure to observe artistic dictates on his frustration at having to
complete with such nonsense:

It's true that I've sometimes written
following the art known to few,
but when I see those monstrosities
full of special effects come on the scene,
attracting the throng of women and the rabble,
who canonize this pitiful exercise,
I go back to that barbarous habit,
and when I'm going to write a play,
I lock up the precepts with six keys. . .

In *Acting Is Believing* Genesius tells Diocletian: "I have a play by a
Greek who bases all his works on lowering monsters from heaven
and raising them back up. The theater has so many drawers and curtains, it looks like a writing desk. There's no chess board like his
canvas. If you look at his verses all together, they seem like stones
sorted out by a rustic while threshing. But they usually amaze the
ignorant mob and bring in more money than good plays, because
they talk nonsense, and even if a couple of people take offense, there
are still over five hundred who like it."

Lope's aversion to the precious language and convoluted syntax
affected by Góngora and his followers is well known. In the *New Art*
he had advised aspiring playwrights:

Don't quote scripture nor offend
the language with esoteric words,
for if one is to imitate real speech,
it can't be with "Panchaias," "Metauros,"
"hippogriffs," "demigods," and "centaurs."

In *Acting Is Believing* he takes the opportunity to satirize the writing
of a certain priest-playwright whom he calls Fabritius. Genesius tells
Diocletian that Fabritius's poetry is "harsh, priestly and precious. If he

has a chance to call the sun 'eternal lamp,' you can bet he won't call it 'Phoebus.' He ransacks the odors or spices of both Indies and manages to find a place for every wild animal and serpent in Libya." The emperor succinctly comments: "They should be his audience."

One can well imagine how Lope would have relished such a conversation with his own monarch, Philip III. Infuriated by the court's preference for other playwrights whom he knew to be his inferiors, Lope must have found great consolation in this story of a playwright who, having attained the height of royal favor, humiliated the emperor by rejecting his generosity and choosing to serve a higher master.

4

While researching his play's historical background, probably in Pero Mexía's *Imperial History,* Lope seems to have been much impressed by the events that led up to Diocletian's election as emperor. In fact Lope found Diocletian's story so attractive that he developed it in *Acting Is Believing* to such an extent that it almost overwhelms the story of St. Genesius. Genesius makes only one brief appearance in the play's first act, which is otherwise entirely devoted to the Diocletian story, and Diocletian is on stage during all but about ten minutes of Act II. It is in fact arguable that Diocletian is the real protagonist of the play, though it is probably more accurate to say that he and Genesius are equally important.

The most interesting passage in Mexía's chapter on Diocletian reads as follows: "Dioclesianus being a poor soldier and in his youth serving in the wars of France, was lodged in a woman's house which was a magician or soothsayer or, as we say, a cunning-woman, one of those which by the Frenchmen were called Druids, which woman was his hostess and did dress his meat. And making his reckoning with her for his diet (which he did every day), as he was very sparing of his purse, this woman said unto him: 'Thou are very sparing with me, Diocletian, and truly art too very a niggard'. Diocletian in jest answered her: 'Pardon me that I am so sparing in my expenses, for I am but a poor soldier; but I promise thee that when I am an emperor, I will be very liberal unto thee'. Quoth this woman unto him: 'Do not jest with me, for thou shalt be emperor when thou hast slain a boar.' "[1] One of Lope's favorite dramatic devices was to begin a play with a foreshadowing of later developments, and this well-known legend offered him an ideal starting point for *Acting Is Believing.* In Lope's play, however, the "soothsayer" becomes an attractive young woman whom Diocletian will reward for her prophecy by making her his wife. The audience has hardly had time to digest this intriguing episode when a violent storm erupts, and the Emperor Aurelius bursts on stage shouting his defiance to heaven: "Don't you see that my sacred laurel is proof against your fury? ... I am Aurelius Carus, I am the Roman Caesar. I sustain the lower world with my protection. If the firmament is yours, the earth is mine, for thus the power is divided." His arrogant words are quickly punished when he is struck down by lightning.

Lope's audience was well aware that it was not only in ancient Rome that kings had thought themselves invulnerable and semi-divine. A scant fourteen years before Lope wrote *Acting Is Believing,* Father Pedro de Rivadeneira had addressed the future Philip III in a treatise on the Christian Prince in the following terms: "The being and power of the king is a participation in the divine being and power... The good king is like another sun in the world and a god on earth, whose vicar and minister he is; and people look upon him and respect him as they do God, whom he represents by watching out for and preserving the common good, as God does."[2] These words did not fall on deaf ears. The modern editor of Rivadeneira's work, Vicente de la Fuente, tells us that Philip "read and re-read the book in the midst of his habitual idleness, soaking up its teachings..."[3] Rivadeneira's *Treatise* offers a theological basis for the theory of the divine right of kings. The most important aspects of that theory, all of which Rivadeneira defends not as theory but as divinely revealed truth, have been succinctly summarized by John Neville Figgis in the following four propositions: "(1) Monarchy is a divinely ordained institution; (2) Hereditary right is indefeasible; (3) Kings are accountable to God alone; (4) Non-resistance and passive obedience are enjoined by God."[4] But although the theory of the divine right of kings had begun to win adherents in Spain as early as the mid-fifteenth century and by Lope's time was almost universally accepted, some dissenting voices were raised. In his *De Legibus ac Deo legislatore* Francisco Suárez argued that power always resides in the community, which conditionally confers the right to exercise it on the king. The theory of the monarchy as social contract was much more fully developed in still another treatise addressed to Philip III, Father Juan de Mariana's *De rege et regis institutione* (1599), in which Mariana undertook to define the limits and controls on royal power. It is surely significant that, while the treatises that supported the divine right of kings were published in Spanish and widely read, those that denied or questioned the doctrine were written in Latin and their readership was limited to a select audience of scholars.

Consequently, as José Antonio Maravall has pointed out, the influence of the latter works "was in reality very slight, though it may indeed have served as a distant ideological counterpoint."[5]

Perhaps the most radical aspect of Mariana's treatise, and certainly the one which brought it the most notoriety, was his argument that it is lawful to overthrow and even assassinate a tyrant. Mariana devotes the fifth chapter of his work to an analysis of the differences between the good king and the tyrant. The following, in his view, are the outstanding characteristics of the good king: "It is proper for a good king to defend innocence, repress wickedness, save those who are in danger and procure happiness and every kind of good for the republic . . . The king is, furthermore, humble, courteous, accessible and disposed to live under the same law as his fellow citizens . . . The king exercises the power he has received from his subjects with singular temperance . . . He does not exclude the poor or the homeless from his palace nor even from his chamber. He lends an attentive ear to the complaints of all. He does not allow cruelty or harshness to be practiced in any part of his kingdom. He doesn't rule his subjects as slaves but rather governs them as children, realizing that he has received his power from the hands of the people. He aims above all for their love, and only aspires to make himself popular through licit means, deserving the benevolence and applause of his vassals, especially of the good ones. He must never consider himself owner of the republic nor of his vassals, no matter how much flatterers may insinuate that this is the case. He must believe that as head of state, he is granted a certain allowance determined by the citizens, an allowance that he will never dare increase unless the people themselves so resolve. However, this does not relieve him of the responsibility of accumulating wealth or increasing the public treasury, which he will place in the most brilliant condition without causing his subjects to shed a single tear. He will do this by means of the spoils he conquers from his enemies. He will live modestly in his palace and will accommodate his expenditures to the income from taxes, always making sure that this is sufficient, whether to maintain the peace or to carry on warfare. The king will be that much better able to serve as a model for his fellow citizens if he succeeds in banishing flatterers from his palace, for they are very pernicious men. It seems to me, finally, that all the deeds of princes should be destined to foment benevolence in the breasts of their subjects, ensuring that the latter always live under their government with the greatest possible happiness."[6]

The tyrant, on the other hand, "believes that his greatest power lies in being able to give free rein to his passions. He considers no wickedness unseemly, commits all kinds of crimes, destroys the wealth of the powerful, violates chastity, slays the good and comes to the end of his life without having neglected to perform a single vile action."[7] In the following chapter Mariana states that if a tyrant refuses to respond satisfactorily to advice or warnings, one should cease to recognize him as king – since his royal status was contingent upon the goodwill of his subjects, which he had forfeited – and proceed to take whatever action might be necessary, including assassination, to dethrone him. Mariana is aware that some scholars, though they may agree with him in principle, will object to his expounding his views in public for fear that they will lead to a proliferation of political assassinations and a chaotic situation. He realistically argues, however, that few men will be so selflessly devoted to the cause of liberty that they will be willing to risk life and limb in order to rid the nation of a tyrant. History, he says, records that there have been many monstrous tyrants, but very few of those have been assassinated. Nevertheless, "it is always worthwhile for princes to be convinced that if they oppress the republic, if they make themselves intolerable by their vices and crimes, they are subject to assassination, not only by law but even with the approval and glory of future generations. This fear will at least keep them from surrendering so easily or so completely to lust and to the desires of their corrupt courtiers, so that their fury will be restrained, at least for a while. This fear will be able to deter them a good deal, and even more than this fear, the knowledge that the authority of the people is always greater than their own, no matter how much evil men, in an attempt to flatter them, may affirm the contrary."[8]

I am convinced that Lope was familiar with the passages I have just cited, and furthermore, that he agreed with them wholeheartedly. The characters Diocletian and Carinus in *Acting Is Believing* perfectly embody the traits of the good king and the tyrant, as outlined by Mariana. We know of Carinus by reputation before he makes his initial appearance on stage. In the very first speech in the play, the soldier Marcellus complains that Aurelius is leading his men in a senseless campaign against the Persians, while his son Carinus has become "a satyr at Hippodamia's wedding." Shortly thereafter, Marcellus's comrade Maximian exclaims: "It drives me mad to hear about Carinus. They say he lives a bawdy life in Rome, raping the most honorable women, sparing neither senators nor Vestal Virgins. A thousand nobles he has killed and brought to shame without respect for hoary-headed consuls, judges, praetors, those who oft have triumphed." When Carinus finally appears, he has disguised himself as a nobleman to go out for a night of revelry with his mistress Rosarda, his servant Celius and two musicians. When Celius tells Carinus that kingship is only a role, which death will ultimately strip from him, Carinus haughtily replies: "What's all this about death and wearing the king's costume? I'm the Roman Caesar, sovereign lord, not some make-believe emperor...What is death? What kind of nonsense is it to say that a king dies? Human laws cannot affect the Emperor Carinus." Rosarda warns Celius that Carinus expects "pleasure, flattery, and lies," not truth, from him. Nevertheless, Celius persists in admonishing Carinus that the city is in an uproar over his outrageous behavior, and advises him: "Reform and learn your lesson." The warning comes too late, however, for at that moment the consul Laelius, whose wife Carinus had raped, arrives determined to kill the tyrant. When Carinus accuses Laelius of the crime of *lèse majesté,* Laelius tells him: "you forfeited your majesty when you sinned against honor. You, who are obliged by that sacred laurel which you wear on your brow, to defend all honor, instead robbed me of my honor, and at that moment you were bereft of honor yourself and lost your royal status. You are no king. No one can be king unless he rules in his people's hearts. No one who has offended heaven with so many crimes can be king. What have you done since you took office? What gold have you added to the treasury, since you've spent what was there on madmen and whores? What provinces have you added to the Roman people?...What virtuous man have you rewarded, as your great father Aurelius and your brother Numerianus did? Your rewards have gone instead to rascals, pimps and scoundrels." Laelius then stabs Carinus, who acknowledges as he is dying that "I played my part. I was Caesar, I was king, I was Rome. The tragedy is over, death has stripped me naked. I suspect that my whole life hasn't lasted an hour and a half. Since the law is the same for all who are born, put these trappings of an actor-king where my successor can take them up again, for he too will have to play his part, heaven grant with more success than I!"

It is hard to imagine a more exact dramatic adaptation of Mariana's doctrines than this scene. Carinus, who corresponds exactly to Mariana's description of the tyrant, is shown to have forfeited his royal status by failing to fulfill his duties as monarch and is justly executed by Laelius in punishment for his crimes. The radical implications of a similar assassination of a tyrant by an outraged populace in *Fuenteovejuna* have often been dismissed by the argument that the Catholic monarchs attempted to punish the "guilty" parties and finally permitted the deed to go unpunished only because they were unable to isolate the ringleaders and because the tyrant in question had taken up arms against them in a civil war. Hence, we are told, the play does not really condone tyrannicide. No such argument can be made in the case of *Acting Is Believing*. Near the end of Act I we are told that Laelius remained in Rome, where he was protected by a certain "Marcus Octavius" and that the people welcomed their new emperor "with pleasure."

If Carinus embodies the characteristics of the tyrant, Diocletian is an exact representation of the good king. He is devoid of greed or self-interest. When the army elects him emperor, he gives them the spoils of Aper and Numerianus's tent, saying: "I'm satisfied with this clothing and this sword to defend you." Upon arriving in Rome, he pays the soldiers and distributes money among the people, stating that "the only thing I want to acquire and store up is goodwill." He invites the senators and their wives to dine with him and provides gladiatorial games for the people. He rewards his friend Maximian's loyalty by making him his coadjutor. Camilla, the peasant woman who had jokingly predicted that he would become emperor after slaying a wild boar, is admitted to the palace and allowed access to Diocletian at all times. In his treatment of Genesius, he shows himself to be a discriminating and generous patron of the arts. His response to the flattery of his courtiers is a model of humility. He chooses a consort worthy of his love and is faithful to her. Even his decision to have Genesius put on a play ridiculing Christianity is taken in defense of Roman religion, "just to ridicule those who deny their rightful incense to Mars, Venus, Jupiter, and Mercury." When Genesius converts to Christianity, attacks the Roman religion, and insults the emperor, Diocletian shows remarkable restraint, hesitating to believe that Genesius means what he is saying. Even when Diocletian exiles Genesius's company from Rome, the actor Octavius says: "Beloved Rome, I don't blame Caesar nor those who bear his maces, who have indeed been equal to him in mercy. I blame Genesius, who chose to perform the end of his life just at the moment of our richest opportunity," and the audience is almost inclined to agree with him.

Not content with attacking the divine right of kings, the play even calls into question the very principle of the hereditary monarchy. In a song preceding the play-within-the-play in Act II, we are shocked to hear that

> Princes, of great kings the seed,
> Too oft their heritage do squander;

'Tis only virtuous word and deed
Can win them praise both here and yonder.
Diocletian earned his crown
By his deeds of heav'nly splendor.
Now spread abroad his just renown
And to him your homage render.

Admittedly, *Acting Is Believing* is set in ancient Rome, not contemporary Spain, and is based to a large extent on history. Would Lope's audience therefore have failed to see its relevance to their own situation? I think not. Philip III was no lecher, but all the other charges leveled against Carinus in the play applied equally well to Philip and to his henchman, the Duke of Lerma. Philip had debased the currency and emptied the treasury, wasting huge sums on luxury and display. The court expenditures, which in the time of Philip II had totaled approximately 400,000 ducats per annum, by 1607 had reached the shocking figure of 1,300,000 ducats.[9] The situation had become so serious that the government was forced to declare bankruptcy in 1607. At the same time, Philip was engaged in an expensive and unpopular war in Holland. The soldiers there, who waited years for their pay, had several times risen in revolt against their commander, General Ambrosio Spínola. Lope's play opens with a scene in which the Roman soldiers stationed in Persia, lacking money even to buy food, rhetorically ask the emperor: "You, who command so many fierce and warlike men, why don't you govern and sustain them?" In spite of the king's much-touted piety, morals in the court had sunk to an all-time low. For years Lerma and his friends had been shamelessly stealing and misappropriating public funds. A major scandal erupted in 1607 when Pedro de Franqueza, an intimate of Lerma's and a high-ranking court official, was arrested for embezzlement and for having accepted bribes. Noblemen, many of whom had been among Philip II's most trusted advisors and who had served the government long and faithfully, protested against these abuses and were promptly punished by being stripped of their positions at court, banished, imprisoned, or even murdered. It was in this atmosphere that Lope wrote *Acting Is Believing.*

We can assume that some of the venom Lope unleashes against the king in the play resulted from his own resentment at not having received the recognition he felt he deserved from the monarch. I think it is only fair, however, to attribute much of the political criticism in the play to Lope's outraged patriotism.

Acting Is Believing belies the often-voiced charge that Lope de Vega was "a paradigm of what we would today call reactionary..."[10] As Charlotte Stern has reminded us in an important recent article, "if Lope's plays strike us as propaganda, it is because very few have been subjected to the kind of scrutiny designed to uncover their uniqueness."[11] She argues persuasively that Lope's plays frequently embody the tensions and contradictions inherent in his society and compel the spectators to examine their social norms critically. Like Juan de Mariana, Lope respected the institution of the monarchy, considering it the best possible form of government, but he was able to distinguish between the institution and the individual, between the role and the actor, between make-believe and reality.

5

The original Spanish version of *Acting Is Believing* is of course entirely in verse. Lope employed a total of ten different verse forms in the play, but almost ninety percent of the text is in octosyllabic verse, the meter which most closely approximates the rhythm of normal Spanish prose. More than one-fifth of the play's verse rhymes in assonance, which is much less obvious than consonant rhyme. These facts, combined with Lope's celebrated facility for versification, produce the effect of a very unobtrusive verse, which lends subtle rhythms and nuances to the work but in recitation is often scarcely distinguishable from prose. I initially experimented with translating the play into English blank verse, but I found that my translation soon began to take on a monotony which was in sharp contrast to the spontaneity and variety of the original. I therefore decided that the only way I could retain those important qualities was to translate the play in prose. Where it seemed appropriate, I have tried to duplicate the rhythmic nature and sonorousness of the original. The two places where I found that prose would simply not do were in the play's sonnets and songs. The sonnets are highly artificial and lyrical in the original, and only a verse translation could do them justice. The songs, if they are to be set to music—the original settings are probably irretrievably lost—had to be in verse. I have also attempted to echo the somewhat archaic flavor of Lope's seventeenth-century Spanish, but without succumbing to the temptation of creating a belabored pastiche of Elizabethan English. Lope wrote in the language of his people, and I have tried to follow his example in this translation.

I am grateful to Professor Victor Dixon of the University of Dublin for reading a first draft of the translation and pointing out a number of errors and omissions. Professor Bruce Wardropper of Duke University has given me invaluable moral support and encouragement in this project and has also helped me to find suitable English equivalents for a couple of seemingly untranslatable puns. Dean David Gitlitz of the State University of New York at Binghamton has also offered helpful suggestions.

Notes

1. I have used W. Traheron's seventeenth-century English translation of Mexía's work, *The imperiall historie, or The lives of the emperors, from Ivlivs Caesar, the first fovnder of the Roman monarchy, vnto this present yeere* (London: Printed by H. L. for M. Lownes, 1623), p. 217.

2. *Obras escogidas del Padre Pedro de Rivadeneira,* ed. Vicente de la Fuente, Biblioteca de Autores Españoles, No. 60 (Madrid: Sucesores de Hernando, 1919), p. 452.

3. Ibid., editor's introduction, p. 449.

4. *The Divine Right of Kings* (Gloucester, Mass.: Peter Smith, 1970), pp. 5-6.

5. *Estado moderno y mentalidad social* (Madrid: Revista de Occidente, 1972), I, 264.

6. *Obras del Padre Juan de Mariana,* Biblioteca de Autores Españoles, No. 31 (Madrid: Rivadeneyra, 1854), pp. 477-78.

7. Ibid., p. 477.

8. Ibid., p. 483.

9. Ciriaco Pérez Bustamante, *La España de Felipe III,* in *Historia de España dirigida por Ramón Menéndez Pidal,* vol. 24 (Madrid: Espasa-Calpe, 1979), 437-38.

10. Carlos Blanco Aguinaga, Julio Rodríguez Puértolas, and Iris M. Zavala, *Historia social de la literatura española* (Madrid: Castalia, 1978), I, 36.

11. "Lope de Vega, Propagandist?" *Bulletin of the Comediantes,* 34 (Summer 1982), 24.

Bibliography

This list includes only sources I have actually consulted in preparing my translation and the introductory essay.

I. Editions of *Lo fingido verdadero*

Decima sexta Parte de las Comedias de Lope de Vega Carpio. Madrid: Viuda de Alonso Martín, 1621. *Lo fingido verdadero* is the last of the twelve plays in this volume.

Obras de Lope de Vega, IV. Ed. Marcelino Menéndez y Pelayo. Madrid: Real Academia Española, 1894. *Lo fingido verdadero* occupies pp. 43-79 of the volume, and a critical introduction to the play appears on pp. XXXIX-LIII. This edition has been reprinted in Vol. 177 of the Biblioteca de Autores Españoles (Madrid: Atlas, 1964).

II. Editions of other works by Lope de Vega

La Arcadia. Ed. Edwin S. Morby. Madrid: Clásicos Castellanos, 1975.

El arte nuevo de hacer comedias en este tiempo. Ed. Juana de José Prades. Madrid: Consejo Superior de Investigaciones Científicas, 1971.

Colección escogida de obras no dramáticas. Ed. Cayetano Rosell. Biblioteca de Autores Españoles, No. 38. Madrid: Atlas, 1950.

La Dorotea. Ed. Edwin S. Morby. Berkeley: University of California Press, 1968.

La Dorotea. Trans. and ed. Alan S. Trueblood and Edwin Honig. Cambridge, Mass.: Harvard University Press, 1985.

Epistolario de Lope de Vega Carpio. Ed. Agustín G. de Amezúa. 4 vols. Madrid: Real Academia Española, 1935-1943.

Obras escogidas. Ed. Federico Carlos Sainz de Robles. Vol. II: poesías líricas, poemas, prosa, novelas. Madrid: Aguilar, 1973.

El peregrino en su patria. Ed. Juan Bautista Avalle-Arce. Madrid: Clásicos Castalia, 1973.

III. Books and articles on Lope de Vega's life and works

Barrera, Cayetano Alberto de la. *Nueva biografía de Lope de Vega.* Biblioteca de Autores Españoles, Nos. 262-263. Madrid: Atlas, 1973.

Castro, Américo, and Hugo A. Rennert. *Vida de Lope de Vega.* Salamanca: Anaya, 1968.

McCready, Warren T. "Lope de Vega's Birth Date and Horoscope." *Hispanic Review,* 28 (1960), 313-319.

Menéndez Pidal, Ramón. *De Cervantes y Lope de Vega.* Colección Austral, No. 120. Madrid: Espasa-Calpe, 1964.

Montalván, Juan Pérez de. *Fama póstuma a la vida y muerte del doctor Frey Lope Félix de Vega Carpio.* In *Comedias escogidas de Frey Lope Félix de Vega Carpio,* ed. Juan Eugenio Hartzenbusch. Biblioteca de Autores Españoles, No. 24. Madrid: Rivadeneyra, 1853.

Montesinos, José F. *Estudios sobre Lope de Vega.* Salamanca: Anaya, 1967.

Morley, S. Griswold, and Courtney Bruerton. *Cronología de las comedias de Lope de Vega.* Trans. María Rosa Cartes. Madrid: Gredos, 1968.

Stern, Charlotte. "Lope de Vega, Propagandist?" *Bulletin of the Comediantes,* 34 (Summer 1982), 1-36.

Trueblood, Alan S. *Experience and Artistic Expression in Lope de Vega: The Making of "La Dorotea."* Cambridge: Harvard University Press, 1974.

Trueblood, Alan S. "Rôle-Playing and the Sense of Illusion in Lope de Vega." *Hispanic Review,* 32 (1964), 305-318.

Vossler, Karl. *Lope de Vega y su tiempo.* Trans. Ramón de la Serna. Madrid: Revista de Occidente, 1933.

Lo fingido verdadero

Acting Is Believing

Translator's Recommendations to Directors

The stage should have a stark appearance, with lights and technical equipment exposed to view. The rear of the stage is arranged as a dressing area, with mirrors, costumes hanging on racks, and the usual dressing room clutter. Above the dressing area is a platform large enough to accommodate four people, which can be reached by a crude wooden staircase. A simple blue curtain can be drawn to conceal both the dressing area and the platform. As the audience enters the theater, the actors can be seen dressing, putting on makeup, and talking among themselves. Then the lights are lowered, the blue curtain is drawn, and the play begins.

There are approximately forty parts in the play, but since the theater companies of the time consisted of about sixteen to twenty players, we may be sure that most of the actors played more than one part in the play. The text of Lo fingido verdadero, as it has come down to us, gives us some important clues regarding which roles were played by a given actor. One of the soldiers who appears in the play's first scene is originally listed as "Marcio." In the stage directions to scene two, however, he is listed as "Marcelo." He has no lines in that scene. In scene four he reappears as "Marcelo," not only in the stage directions but in the attributions. In Act III, scene 6, "Marcio" appears as a member of Genesius's theater company. I have deduced that the same actor played Lelio and Léntulo, because in Act II, scene 4, a series of speeches is falsely attributed to Lelio, when Léntulo is actually the character on stage. The graph on the following page indicates how all parts in this play could be covered by a minimum of eighteen actors.

It should be emphasized, however, that the doubling of parts in this play is not just a matter of convenience or even necessity. Lope uses it as a device to underline the fact that everyone is called upon to play multiple roles in his or her life. Thus, even though the same actor plays Diocletian throughout the play, his role changes from that of a poor soldier who was the son of a slave to that of Roman Emperor. Similar changes take place in the roles of Camilla and Maximian. Hence, no elaborate efforts should be made to conceal the fact that each actor is playing several parts. They might even be shown discarding one costume and putting on another. If some confusion results, that seems to be part of the playwright's intent. The scene divisions in the translation are my own. I have indicated these divisions by bracketed numerals in the page margins. Most of the divisions occur when the stage is emptied of actors,

	Act I				Act II					Act III							
Scene	1	2	3	4	1	2	3	4	5	1	2	3	4	5	6	7	8
1. Curtius		X		X									Sulpicius	X	X	X	
2. Marcellus		X		X			Patricius			Rutilius				Captain	Martius	X	
3. Maximian		X		X	X		X	X		X			X	X			
4. Diocletian		X		X	X		X	X		X			X	X			
5. Camilla		X			X			X		X			X	X	Celia		X
6.		Aurelius Caesar										Offstage voice		Angel #1		Warden	
7.		Numerianus		X				Singer				Fabius		Singer/Angel #2 Fabius	Fabius		X
8.		Aper		X		Pinabellus		X						Soldier #1	Salustius		X
9.			Carinus			Octavius		X	X		X			Soldier #2	Octavius		X
10.			Celius		Servant			Celius						Soldier #3	Celius		
11.			Rosarda					Marcella			X			X	X		X
12.			Musician #1	Soldier #1	Courtier			Musician #1						X	Albinus		X
13.			Musician #2	Soldier #2	Courtier			Musician #2						X	Ribete		X
14.			Genesius		X	X		X	X	X	X	X		X	X	X	X
15.			Laelius		Lentulus	X	X	X					X	X	X	X	
16.			Fabritius					Fabritius						Singer	Fabritius		X
17.			Man #1	Severus	Courtier									Angel #3	Sergestus		X
18.			Man #2	Felisardus	Courtier			Singer						Angel #4			

and a change of setting is usually also implied. There are, however, three cases where Genesius remains alone on stage for a soliloquy before being joined by another actor (Act II, scene 2; Act III, scene 2; and Act III, scene 3). In each of these cases an important change of mood occurs, reflected in the Spanish original by a change of versification. I have also chosen to set the two plays-within-the-play (Act II, scene 3, and Act III, scene 5) apart as separate scenes.

M. D. M.

Lo fingido verdadero

Characters

MARCELLUS, a soldier
CURIUS, a soldier
MAXIMIAN, a soldier and, later, co-emperor with
DIOCLETIAN
CAMILLA, a peasant and, later, Diocletian's wife
AURELIUS CARUS, Emperor of Rome
NUMERIANUS, Aurelius' son
APER, Consul and Numerianus' father-in-law
CARINUS, Aurelius' son and co-emperor
ROSARDA, Carinus' mistress
CELIUS, Carinus' servant
GENESIUS, an actor and playwright
LAELIUS, a Consul
SEVERIUS, an army officer
FELISARDUS, a Roman official
LENTULUS, a senator
PINABELLUS, an actor who plays servants
PATRICIUS, a senator
MARCELLA, an actress who plays leading ladies
FABRITIUS, Marcella's father; an actor who plays old men
OCTAVIUS, an actor who plays leading men
CELIUS, an actor
RUTILIUS, a palace servant
FABIUS, a boy actor
SERGESTUS, a Roman official
ALBINUS, a comic actor
SALUSTIUS, an actor who plays villains
CELIA, a supporting actress
RIBETE, a property man
MARTIUS, a stage hand
A WARDEN
SINGERS
MUSICIANS
SOLDIERS
GUARDS
COURTIERS
ANGELS

Act One

[Enter Marcellus] [I.1]

MARCELLUS.   Damn the emperor and all empires! Dragging us like this through Mesopotamia, ragged, naked, pelted with reproach, in search of infamy instead of glory! He's become another Nero in our hateful eyes, and his son in Rome, a satyr at Hippodamia's wedding, while we slog on from sea to sea to bring the Persians to heel.

[Enter Curius]

CURIUS.   Does this make sense? Caesar knows that his son Carinus, infamy of Roman valor, lives a life of madness and excess, yet he leads us to conquer the Persians. His mighty brother, Caesar Numerianus, should be raised to the altars as a god, but instead, he sends him to his death in the merciless July heat, just to show the Persians our iron will.

[Enter Maximian]

MAXIMIAN.   If Numerianus were less good, no soldier would remain, Aurelius Carus, to follow your standards and your eagles; so much do they esteem his leadership. Just when the army thought your churlish heart had softened and you would take us home, instead, in burning July, you lead us against the Persian rebels. Soldiers, will you stand for this?

[Enter Diocletian]

DIOCLETIAN.   By holy Jupiter, whom Mt. Olympus worships, if there were two others who agreed with me, no soldier, Aurelius, would follow you! Now you lead us towards the dawn, when the sun shines in Virgo and fiery Leo feels its rays more than the strength of valiant Hercules. Have you perchance a brain? Have you a soul? After all the voyages we've made with you, sometimes in tempest and sometimes on calm seas, as far as the sources of the sacred Jordan, you now propose to lead us through the palms to Tigris's sandy banks and the frozen Cydnos, where the unvanquished Alexander met a brutal death? Maximian!

MAXIMIAN.   Brave Diocletian!

DIOCLETIAN.   Marcellus, Curius, what is this?

CURIUS.   We're calling to account this French Emperor of Rome, who's taking us off to the realm of Semiramis.

DIOCLETIAN.   Well may he thank Numerianus for protecting his freedom when he might instead have seized the name of Caesar.

MAXIMIAN.   It drives me mad to hear about Carinus. They say he leads a bawdy life in Rome, raping the most honorable women, sparing neither senators nor Vestal Virgins. A thousand nobles he has killed and brought to shame without respect for hoary-headed consuls, judges, praetors, and those who oft have triumphed.

MARCELLUS.   Why didn't he put that bloodthirsty beast's brother Numerianus in Rome? He's so learned in all letters, so humane! Then the empire would again be blessed. Our homeland, famous for such Caesars, once again would see the happy age of Trajan's reign.

DIOCLETIAN.   Because he does nothing right, and because arrogance dominates the Romans with such monsters. Is there anything to eat?

CURIUS.   You ask for food?

DIOCLETIAN.   No money either?

MAXIMIAN.   Money! If we had money, I suspect I would suffer the labors or fables of Hercules and many more.

DIOCLETIAN.   I'd not be Caesar. You, who command so many fierce and warlike men, why don't you govern and sustain them? Help me fling insults at Aurelius!

MAXIMIAN.   He's an old fool!

CURIUS.   A drunkard!

MARCELLUS.   A madman!

DIOCLETIAN.   He's the emperor. Let's watch our tongues; we mustn't disrespect the royal sceptre, even if wielded by a barbarous Circassian.

CURIUS.   Wasn't it you who provoked us?

DIOCLETIAN.   If I did, I'm sorry for it.

MARCELLUS.   It doesn't matter.

DIOCLETIAN.   I tell you this because I may yet be emperor.

CURIUS.   Who, you?

DIOCLETIAN.   Yes, I, someday.

[Enter Camilla, with a basket of rolls]

CAMILLA. Who'll buy my good bread, soldiers? It's white and well done.

MAXIMIAN. Victuallers here so soon? We're not so badly off.

DIOCLETIAN. We had no reason to complain of Caesar.

MARCELLUS. Yes, but as for money...

MAXIMIAN. Today I hope for help.

DIOCLETIAN. Calm down, and praise Caesar once again.

CURIUS. When I've eaten.

CAMILLA. Who'll buy my bread, who'll buy it?

MAXIMIAN. I'll buy.

MARCELLUS. And so will I.

DIOCLETIAN. Camilla, God bless you, give me a roll, since you know me.

CAMILLA. Who'll give me money?

DIOCLETIAN. I can't just now, but I'll pay you when I'm Emperor of Rome.

CAMILLA. I like your sense of humor.

DIOCLETIAN. And why not?

CAMILLA. How should I know? But this is not the first time, wretched soldier, that you've promised you would pay what your necessity requires of me when you don the emperor's laurel crown. You never change.

DIOCLETIAN. Is Rome a laughing matter? I have only to get there to be her emperor and Caesar, lord of all the world with power absolute.

CAMILLA. And is a roll a laughing matter, and taking it for nothing?

DIOCLETIAN. For God's sake, woman, I expect to pay you when they give me the empire.

CAMILLA. Here you are, and you'll be emperor when you kill a wild boar.

MAXIMIAN. If it's that easy, take it; don't hesitate.

DIOCLETIAN. No, but it gives me pause. This woman has just told me that I'll be Caesar when I kill a wild boar. I've taken the omen so much to heart that, even if it was a joke, I expect it to come true, for though I'm but a lowly soldier, I've killed a thousand wild boars.

CURIUS.   Never trust in omens.

DIOCLETIAN.   The whole Roman religion rests on omens.

CURIUS.   Eat, and cease to think of what can never happen.

DIOCLETIAN.   Take the bread, comrades, and if by some mystery I
    one day see myself as emperor, I'll know how to share the
    empire as I share this bread. Take and eat, Maximian; you'll be
    co-emperor with me. And you, my dear *[to Camilla],* if I
    become a Roman Caesar, you'll see what I'll do for you.

CAMILLA.   I love your spirit! Do you want all the bread now and as
    much as I can bring? Would you like the money too? Here, take
    it.

DIOCLETIAN.   When I am emperor, I'll pay you for this love in
    Rome.

CAMILLA.   Though I am humble, I can't resist a man who thinks
    big.

CURIUS.   What raging winds!

CAMILLA.   It's going to rain. I'm leaving, but first, tell me your
    name.

DIOCLETIAN.   My name is Diocletian.

CAMILLA.   And when you kill a wild boar, you'll be Rome's Caesar.

[Exit Camilla]

MAXIMIAN.   The sky is growing darker.

MARCELLUS.   What terrible darkness!

DIOCLETIAN.   It looks to be a frightful night. The thunder and the
    lightning is like a battle's raging.

CURIUS.   The cloud's black breasts burst open, shooting out fire
    that lights up the sky.

DIOCLETIAN.   They're like silver serpents that split their element.
    Even here you can hear the awful bellowing of the sea.

MARCELLUS.   The united elements are trying to break asunder.

CURIUS.   The wind is ripping the tents and pavilions from the
    ground.

DIOCLETIAN.   Why shouldn't it tear out a dry stake tied to four
    ropes, since it's uprooting ancient oaks, lush hedges, savins,
    strong, robust ilexes, palms, and noble laurels from their native
    soil?

CURIUS.   Let's run to that overhanging rock for safety.

MARCELLUS.   Oh, to be in Damascus under a solid roof.

MAXIMIAN.   I'm frightened of the lightning. Run!

DIOCLETIAN.   Don't be afraid. When have you ever seen a Roman
Emperor struck dead by lightning?

[Exeunt omnes. Enter Aurelius Caesar wearing a laurel wreath]   [I.2]

AURELIUS.   What madness is this, heaven, to oppose my angry
arm, after I've crossed mountains of ice and fiery volcanoes?
Don't you see that your hail and lightning are useless against
my strength? You dare spit fierce vipers of fire and snowy bul-
lets when Rome makes bold against the greatness of the Greek
empire? Don't you see that my sacred laurel is proof against
your fury? Don't you see, holy Jupiter, that the eagles of Rome
are yours? This sacred mantel wrapped round my arm and this
naked steel have tamed the Parthian, the Mede, and the region
the sun fears to look upon. I am Aurelius Carus, I am the
Roman Caesar. I sustain the lower world with my protection. If
the firmament is yours, the earth is mine, for thus the power is
divided. And you, holy Mars, since your sons were the first to
sacrifice wild bulls to you in the Roman forum, warming your
altars with blood that stained the sacred rods, how can you
allow storms to divide our army in this wilderness—Rome's
banners and the crown, feared wherever the sun shines from
the Antarctic to Calixto? Jupiter, if I thought the clouds could
show this fearful face without your consent, I would suppose
that you didn't know they were breaking their breasts, pregnant with lightning and thunder, against Rome. You, oh heavens, dare shatter the Roman army! What is this? If you dare
disturb the order wrought by her consuls, then don't be
shocked if the giants decide to make war against you. As Jupiter lives, if we have to pile mountain upon mountain, I'll give
you a fright. Then you can strike me down like Typhon, once
my vengeance has tempered your fury. I'll raise up a thousand
squadrons of fierce men to your exalted sphere. *[He is struck by
lightning]* What, are you punishing my speech? Oh, what a
fierce thunderbolt! Jupiter, you've struck me down: I'm dying.
*[He falls to the ground. Enter Numerianus, Diocletian, Marcellus,
Curius, and Maximian]*

NUMERIANUS.   Come here, soldiers, come here! It struck right next
to my father's tent.

DIOCLETIAN. Impious heaven, your rage does not respect the sacred laurel.

NUMERIANUS. What?

DIOCLETIAN. He's dead.

NUMERIANUS. Rome, today your crown is undone. It is he, the lightning struck him.

MAXIMIAN. How black his face is!

CURIUS. There's not even a scratch on him.

NUMERIANUS. Dear father, who could look upon you without fainting?

MARCELLUS. Aper, your father-in-law, is coming.

[Enter Aper]

APER. Numerianus, what has happened to Caesar?

NUMERIANUS. Oh, Aper! Can't you see that he's scorched to the bone?

APER. Why are you so bewildered, son?

NUMERIANUS. To see such great prosperity end in so much woe, to see the man who begot me and carried the empire on his shoulders struck down to earth! Like a second Hercules, he sustained the world in his valor, and I thought the world could not go on without him.

APER. My son, lightning strikes high towers. Only lightning could kill so great a person, who supplied Rome's needs and whose brow was honored by a greater crown than that of Augustus. Take hence his body where it may receive due honor, so that from his fire the phoenix Numerianus may emerge with new-born valor.

NUMERIANUS. There's nothing to be done until my brother decides. Carinus rules in Rome. If he chooses me, if he appoints me as Caesar, I will serve the empire, but I doubt that he would share the power even with his own shadow.

DIOCLETIAN. These words are not without mystery.

APER. How do you expect to get the Roman legions to support you?

NUMERIANUS. I won't tolerate insubordination. By imitating the valor of Aurelius Carus against my brother Caesar and taking the title of Consul Numerianus. Ho, brave soldiers! Now you're fighting under my command!

APER.   I trust in sacred heaven and in Jupiter's kindness that they'll gird your brow with laurel. Long live the consul!

ALL.   Long live the consul!

[Exeunt omnes. Enter Carinus with two musicians and Celius, a   [I.3]
servant, and Rosarda, dressed as a man]

CARINUS.   We've played a good trick on her!

ROSARDA.   By God, the emperor fought bravely!

CARINUS.   That's because I wear you on my chest.

ROSARDA.   Is there a piece of armor that has a female name?

CARINUS.   How about "breastplate"?

ROSARDA.   Yes, that's right, but "buckler" sounds masculine.

CELIUS.   Since women are the buck's best friends, you ought to have called her your buckler, sir.

CARINUS.   Your Spanish-style puns are a pain in the ass.

CELIUS.   I can't help it; that's where I'm from.

CARINUS.   Where shall we go now to find a woman alone?

MUSICIAN 1.   That high-spirited Neapolitan woman Falsirena lives near here, but...

CARINUS.   She's a sourpuss.

MUSICIAN 1.   She's old.

CARINUS.   Old, disloyal, and a temptress. Three great qualities! Don't even mention her.

MUSICIAN 2.   A month ago three women came by here, separated like fruit, into three ages: green, ripe, and rotten.

CARINUS.   The green one?

MUSICIAN 2.   Fifteen.

CARINUS.   A nice, juicy age; and the second?

MUSICIAN 2.   Fifteen doubled; fifteen more than the first.

CARINUS.   Thirty's a fair age; and the next one?

MUSICIAN 2.   Twice thirty.

CARINUS.   That's bad: sixty.

MUSICIAN 2.   She's the thirty-year-old's *middle*man.

ROSARDA.   Damn her to hell! Let's stay away from old women.

CARINUS.   Why?

ROSARDA.   They're cranes who spy on the young.

CARINUS.   And who pluck themselves from the chin to the eyebrows.

CELIUS.   Here lives a great lawyer, whose wife is a professional too. She's known for her . . .

CARINUS.   Out with it!

CELIUS.   . . . scrupulous scrutiny.

CARINUS.   I'll bet she's a scrumptious scrutineer, for so I've heard said in a Spanish couplet.

CELIUS.   I once went to see her, and in a single hour she had over a thousand clients.

ROSARDA.   If she gave such scrupulous scrutiny, you shouldn't be surprised.

CELIUS.   Those actors' women are pretty scrumptious themselves.

CARINUS.   Are they married?

CELIUS.   No, sir.

CARINUS.   And could an emperor get involved with those women?

ROSARDA.   You're getting picky! Don't you know they're empresses and queens?

CELIUS.   Do you think that your kingship is better than theirs? The only difference is that their play lasts an hour and a half, while yours lasts all your life. You're an actor too, but you wear the king's costume till death, an ending foreshadowed in the play.

ROSARDA.   How clever!

CARINUS.   You think that's clever? Celius, I've never heard you talk such nonsense in all my life!

CELIUS.   I humbly beg your pardon.

CARINUS.   When I'm out to have a good time, you come up with philosophies about kingship and death.

CELIUS.   It's all just acting. I'm not trying to shorten your days. May Jupiter preserve you a thousand years.

CARINUS.   I'm angry.

ROSARDA.   You're not playing the role of servant very well today, Celius. You're supposed to please, flatter, and lie.

CELIUS.   Then you play the leading lady with the skill only love can give, and talk to angry Caesar.

ROSARDA.   My dear, you see I'm cast as the leading lady in this
   play.

CARINUS.   Having you here is the only thing that makes me feel
   better. Come, let me hug you. What's all this about death and
   wearing the king's costume? I'm the Roman Caesar, sovereign
   lord, not some make-believe emperor. I'm not an actor like the
   man who lives here; I was born king and Caesar to rule and
   command. When Genesius comes on stage to play a king in a
   tragedy, he'll rule for an hour and a half and then will no longer
   be king. But I, who by my good luck am really king, will be
   king in life and in death. My life is secure because of my age
   and health, my strength and valor, and because I'm emperor,
   which is an even greater assurance. What is death? What kind
   of nonsense is it to say that a king dies? Human laws cannot
   touch the Emperor Carinus. It's ridiculous to philosophize
   about this. Human events cannot affect the divine. We emperors, as you know, are almost equal to the gods of heaven; we
   are lords of the world as they are lords of heaven. Damn it, if I
   knew where that rascal Death, who plays scarecrow to the
   world, lived, I'd go there and cut her up! Sing; let there be
   music and freedom while Fortune's wheel is still. Let's make
   some music for Genesius and his actresses.

ROSARDA.   Now you're back in a good mood.

CARINUS.   I have quicksilver on my feet. Let my father march
   through Persia conquering lands for Rome with all his different
   wars; let him call Rome his mother. I'll stay right here and
   enjoy myself at my leisure.

ROSARDA.   Maybe he's already asleep.

CARINUS.   If he's sleeping, he'll wake up.

ROSARDA.   Call for the manager, for I don't think it's Genesius.

CELIUS.   Genesius is just a member of the company, though I think
   he's the best one, for he is also a poet and he writes their plays.

CARINUS.   If he's a bad poet, he should forget about it and not get
   involved in writing them. He can just play his part, for that's no
   mean accomplishment.

CELIUS.   Here he comes.

[Enter Genesius]

GENESIUS.   This house is indeed fortunate, sir, that you should
   come to honor it.

CARINUS. Don't call me by name, for I'm trying to play the part of a simple nobleman. What were you up to?

GENESIUS. I was casting a play.

CARINUS. Who's the author?

GENESIUS. Aristotheles.

CARINUS. A famous wit; it must be wild.

GENESIUS. That it will be, for there's a bull in it. It's the story of Pasiphaë.

CARINUS. Can you say a speech or two for me?

GENESIUS. What for, since you're going to see it?

CARINUS. Do you have music?

GENESIUS. Excellent; and we'll use it if you're interested.

CARINUS. How are you fixed for women? A play is nothing without them.

GENESIUS. I've done the best I could. I had a marvelous one named Lisarda. She was nothing to look at, but she'd bewitch your ears. However, she became a Christian.

CARINUS. She gave up acting.

GENESIUS. She took off right away to become a penitent in the mountains of Marseille.

CARINUS. Would you like me to send for her?

GENESIUS. There's one big obstacle. Another actress here has learned her lines, and I suspect that if I took the part away from her, she'd quit.

CARINUS. Give me a performance right here in the middle of the street.

GENESIUS. I'm afraid, sir, that I won't be able to round up all the actors.

CARINUS. That's true. Some of them are probably out on the town and others must be sleeping.

ROSARDA. I guess it can't be done.

CARINUS. Tomorrow see that a famous poet writes an exciting play about me and Rosarda; but have him portray her as very bright and me as foolish and jealous, and take this moneybag for your trouble.

ROSARDA. Fine. I'll pay him too, so that his famous wit will portray you as loved and an ingrate and me as insanely jealous.

CARINUS. Do you have money on you?

ROSARDA. This golden chain, but it won't do, for it has your portrait on it.

CARINUS. Give it to him, and Genesius will be our official imperial actor, since he has my royal seal which, after all, is my portrait. Portraits were made for absence; there's no need for a portrait when the person is present.

GENESIUS. Let me kiss your feet a thousand times for such an honor and gift.

CARINUS. From now on, you'll be known as Caesar's actor, Genesius. When will you put on the play?

GENESIUS. That depends on the poet to whom I entrust the writing; only he can decide. There are poets whose muses are like women.

CARINUS. Tell me more.

GENESIUS. That's what holds us up, even if you're the one who ordered the play, until she bears fruit, for she takes nine months to bear any play at all.

CARINUS. Yes, but in that case, Genesius, they'll turn out to be sons, and handsome ones at that.

GENESIUS. And what if instead they turn out to be daughters, and ugly ones at that?

CARINUS. Enough! Find Aristeles, and he'll do it quickly.

GENESIUS. He'll observe the rules.

CARINUS. No, do it just like always. I don't like to be limited by art and the precepts.

GENESIUS. The intellectuals will be annoyed.

CARINUS. Well, let them be. Delight the ears, and that's enough, as long as there's no absurdity that can be seen.

ROSARDA. We're wasting the night. Go get some sleep, Genesius.

GENESIUS. With your permission, I'll retire.

[Exit Genesius]

CARINUS. Where shall we go, since my feet are now free? Whenever I'm dealing with poets, I think they're using my feet for their metres.

CELIUS. Let's go and see the cousins.

CARINUS. Don't take me around old women.

MUSICIAN I.   If it weren't so far, there was a wild Spanish woman who would chew you up and spit you out alive.

CARINUS.   I've never seen a Spanish woman who was cold. But it's no fun for me unless we can also make some mischief.

ROSARDA.   What a strange character!

CARINUS.   It really thrills me to steal the honor of a chaste, noble, and virtuous woman, and twice as much, if she's the wife of a senator.

ROSARDA.   Well, I'm telling you that Rome is in an uproar about just that.

CELIUS.   People never stop talking about it, sir. Reform and learn your lesson. Laelius, the consul, whom you've so dishonored, has found out about your crazy love and how you've offended him.

CARINUS.   Celius has turned out to be awfully tiresome tonight!

ROSARDA.   Since you raped his wife, is it any wonder that the Consul Laelius should be angry?

[Enter Laelius with three men]

CARINUS.   You too?

LAELIUS *[to his companions]*.   Stay where you are.

FABRITIUS.   You want to talk to him?

LAELIUS.   Yes.

FABRITIUS.   Are you mad?

LAELIUS.   My honor is at stake. Is Caesar here?

CARINUS.   Who wants him?

LAELIUS.   A consul of your senate, whose wife you raped. But you've done more harm by talking about it than by the wicked deed itself.

CARINUS.   How dare you speak disrespectfully to Caesar!

LAELIUS.   Because you forfeited your majesty when you sinned against honor. You, who are obliged by that sacred laurel which you wear on your brow, to defend all honor, instead robbed me of my honor, and at that moment you were bereft of honor yourself and lost your royal status. You are no king. No one can be king unless he rules in his people's hearts. No one who has offended heaven with so many crimes can be king. What have you done since you took office? What gold have you added to the treasury, since you've spent what was there on madmen and whores? What provinces have you added to the Roman people?

CARINUS.   Have you lost your mind, Laelius?

LAELIUS.   What virtuous man have you rewarded, as your great father Aurelius and your brother Numerianus did? Your rewards have gone instead to rascals, pimps, and scoundrels. Answer me, you monster of vice, viper of Rome, what have you been thinking?

CARINUS.   Heavens, can you permit this insolence without defending me? Rome, am I your master? Is this the kind of son you produce?

LAELIUS.   Just think about how you've behaved, and then try to deny that you deserve my fury. In what tribunal have you passed judgment, in what triumph have you marched, what gift have you offered Rome to make her love and adore you? No, instead you've disguised your majesty to go out at night adorned with a madman and a whore!

CARINUS.   Jupiter be damned! Guards, kill him!

LAELIUS.   What guards? Only two musicians and Rosarda are with you. But I'll free Rome from this new Nero!

[He attacks him, and his companions join in]

CARINUS.   I'm dying! Treason, treason!

LAELIUS *[to Carinus's companions as he tries to make his escape]*.   Let me by, you knaves!

CELIUS.   Is there no help for Caesar?

LAELIUS *[to his companions]*.   Get out of here!

ROSARDA.   Ah, my love!

CELIUS.   What a tragedy!

CARINUS.   I played my part. I was Caesar, I was king, I was Rome. The tragedy is over, death has stripped me naked. I suspect that my whole life hasn't lasted an hour and a half. Since the law is the same for all who are born, put these trappings of an actor-king where my successor can take them up again, for he too will have to play his part, heaven grant with more success than I!

ROSARDA.   Oh, fierce murderous hand! But believe me, when the sacred pyre is lit, love will offer you my life.

CARINUS.   No, Rosarda, live.

CELIUS.   Rome, your Caesar is dead.

MUSICIAN 1.   The whole city is in an uproar.

CARINUS.   Oh, my country, take revenge! Fierce death, where were
you lurking? Who can escape you? What strength is there like
yours, since you slay even kings?

[I.4]   *[Exeunt omnes. Enter Aper with Severius]*

APER.   The illness of Caesar Numerianus forced us, Severius, to
come home, leading the whole Roman army with us. This was
our unanimous decision.

SEVERIUS.   And is Caesar now fully recovered?

APER.   The long journey and the stormy sea have hardly given him
a chance to recuperate. He's receiving the best of care.

SEVERIUS.   Your son-in-law will soon be Emperor of Rome, for his
brother is a barbarian and wants to bring Rome to a sorry end.

APER.   That's where his oppression is leading.

SEVERIUS.   His only interest is vice; he never thinks of the common
good nor even takes a paper in his hands. The people are joy-
fully awaiting Numerianus.

APER.   He's my son-in-law, Severius, and the most gallant prince
this empire has yet seen. I expect great things from his government. It's not without mystery that heaven has preserved him
for this moment.

[Enter Felisardus]

FELISARDUS.   Where is Numerianus?

APER.   Oh, Felisardus!

FELISARDUS.   Oh, brave Aper! Oh, great Severius!

SEVERIUS.   How good to see you! What's new in Rome? How are
the people taking our arrival?

FELISARDUS.   They had prepared triumphal arches for the entry of
the great Aurelius Carus, whose shocking death they hadn't
heard about, but now the arches will be used to honor the fortunate Numerianus. But how sad that on this very day, or rather
night, Carinus went out on the streets in disguise!

APER.   Is he dead?

FELISARDUS.   Laelius, whom he had wronged, killed him.

APER.   In what way did he wrong him?

FELISARDUS. Was it a small offense to rape his wife, a noble matron?

APER. Is your grief just an act?

SEVERIUS. Felisardus is a wise man.

FELISARDUS. Fierce death, what mortal does she spare?

APER. Did Laelius flee?

FELISARDUS. No, for Marcus Octavius is defending him with a thousand men, and the people won't harm him, for they're delighted to call your son-in-law Emperor Augustus.

APER. Go, Severius, and have all the legions you brought home march here; I'll summon all the squadrons as well.

FELISARDUS. An excellent idea, for Rome is burning.

SEVERIUS. While you crown Numerianus with the laurel—may he enjoy it a thousand years in peace—I'll put our men in order.

[Exit Severius]

APER. I need to talk to you in private.

FELISARDUS. You know you can trust me with any problem.

APER. What would the city say if I became Caesar?

FELISARDUS. Just what they've said about many others who were esteemed for their virtue. Are the soldiers fond of you? Have they perchance elected you? Because they have the power to make an emperor, whom Rome, out of fear, will do her best to obey. But as long as your son-in-law Numerianus is alive, I don't think the Roman army will look favorably on your wish.

APER. Alive, you say?

FELISARDUS. Well, isn't he alive?

APER. Pretending he was sick, I've had him carried on a litter to the riverbanks, where the soldiers are welcoming him, although I've kept him out of sight, for, to tell the truth...

FELISARDUS. Go on, there's nothing to be afraid of.

APER. I've killed him, and I've brought his body all wrapped up.

FELISARDUS. You've killed your son-in-law?

APER. It was worth it for so great an empire, for ambition is not the same thing as treason.

FELISARDUS. How do you plan to announce his death?

APER.   In the circumstances, since he was sick anyway, it will be easily believed. And with my great power, reputation, and noble birth, no one else could possibly be elected as long as I'm here.

FELISARDUS *[aside]*.   That Numerianus, so worthy to live that he might restore Rome, should have died!

APER.   What do you say?

FELISARDUS.   That you've done the right thing, and the Roman Senate will approve any Caesar the army chooses, without distinction of birth or reputation.

APER.   What's that?

FELISARDUS.   Let's cut this short, for the soldiers are upset about Carinus's death.

OFFSTAGE VOICES.   Hurrah, hurrah!

APER.   Nonsense; they're all deceived.

OFFSTAGE VOICES.   Long live Numerianus! Hurrah!

APER.   Oh, God! That's the litter.

FELISARDUS.   You needn't worry about their accepting you as Caesar.

[Enter Diocletian, Marcellus, Curius, Maximian, Severius, and two soldiers carrying Numerianus in a curtained sedan chair. One soldier carries a laurel wreath]

DIOCLETIAN.   What difference does it make whether he's sick or not, since the army is dying to see him so that they can give him the laurel crown?

MARCELLUS.   His father-in-law, Aper, is here.

DIOCLETIAN.   Oh, consul, give us permission to do obeisance to Caesar.

APER.   Do as you please; but the only reason I've kept him out of sight is that he was so ill on the trip that I feared for his life. God knows how concerned I've been about his survival.

CURIUS.   Consul, let the army give him the sacred laurel and acknowledge him as its master. The pleasure will give him strength.

APER.   While he's still resting, I'd just like to point out how lovingly I've ruled and guided you since Aurelius died, and that

I've always been like a father to every soldier. Do you recall how much money I've handed out? What poor man have I not helped? Whom have I ever wronged? Nor have I been ungrateful. I give permission for anyone to speak if I've offended him.

MAXIMIAN.   Aper, no one's questioning how good you've always been to the Roman army, but what does that have to do with our wish to do obeisance to Caesar Numerianus? Men, raise those curtains; everyone on your knees.

DIOCLETIAN.   I'll do homage with you.

MARCELLUS.   Give me those divine leaves, for I want to crown him with laurel.

DIOCLETIAN.   His brow does honor to the laurel.

MARCELLUS.   Caesar, all your Roman people are coming to do obeisance to you.

CURIUS.   How pale, how dispirited and sad he looks!

DIOCLETIAN.   Oh, sir! Hasn't being Roman Emperor encouraged and moved you?

MARCELLUS.   Sir, may the virtue of this laurel restore you to health; for I think there's so much power in it that it can give you strength.

DIOCLETIAN.   Observe, great Numerianus, how the whole world offers you its homage; behold how these victorious leaves gird your temples, for though it seems that you do honor to the sacred leaves, they have honored many a brow. He doesn't answer, soldiers. I think he's dead.

MARCELLUS.   Take his hand, and it will tell you if he's alive.

DIOCLETIAN.   Never in the frigid zone have Finmark and Libonia seen such frosty icicles. Caesar is dead.

CURIUS.   What say you?

DIOCLETIAN.   He is dead, and death has turned the crown of evergreen laurels to cypress.

MARCELLUS.   By Jupiter, sickness has not killed him!

MAXIMIAN.   You can be sure of that; it's Aper who has killed him.

CURIUS.   From what he said just now, anyone could see that it was Aper, and that he's given him poison.

APER.   Soldiers, consider well.

MARCELLUS.   Take him away, and we'll give him a mournful bur-
ial with funereal pomp.

[They carry the chair off]

CURIUS.   How uncertain is human glory!

APER.   Soldiers, I am his father-in-law. If Numerianus rests in
peace with the holy gods, he was mortal; that's more than
enough. What you should do now is consider that I alone am
worthy of the green laurel, for neither in the country nor in
Rome is there anyone else who deserves it. It would be
wrong for the army to choose anyone over me, whether he
wears the civilian's toga or the soldier's cloak. I am Aper, I
am consul, I am he whose heroic deeds are known in all of
Asia as well as in Europe. Am I your Caesar, soldiers?

MARCELLUS.   Since you gave poison to your son-in-law
Numerianus, it doesn't seem right to us.

APER.   I, my sons, I?

FELISARDUS.   Aper, why deny your notorious wickedness? You
told me about it yourself. By poison or suffocation, you took
the life of a man who, in all the history of Rome, begging the
pardon of Trajan, that mighty Spaniard, showed the highest
promise of restoring Rome to her rightful place as head and
queen of the world. Would it be fair for you to be crowned
with the laurel that you've robbed from your son-in-law?

APER.   What does that matter, soldiers, as long as my bravery,
my triumphs, and my victories please you?

DIOCLETIAN.   I remember once when a woman was selling bread,
I promised to pay her when I should be Emperor of Rome,
and she told me: "You will be Caesar when your famous
sword kills a wild boar." Heavens! Since the consul's name is
Aper, which in our Latin language means "wild boar," and his
wickedness has turned him into one, surely the time has
come for the prophecy to be fulfilled. But what if the soldiers
get angry and take my life?. . . Still, what does my life matter
when so much is at stake? The soldiers are satisfied and will
accept him as Emperor of Rome. Be still, hand, where are
you going? No one will support me, for I am the son of a
slave. But just to attempt it will suffice to win eternal fame.
If I've fought on land and perilous sea to win fame, and if I've

placed my life, which is short anyway, in certain danger for the sake of an empire and crown which represents the whole world, how could I die with greater honor? Aper, listen.

APER.   Oh, Diocletian! Help me, for if your legion favors me now, tomorrow I'll decorate you with the mural, naval, and siege wreaths.

DIOCLETIAN.   I thank you, consul; but the fearful image of your son-in-law Numerianus, transformed into a dark shade, appeared to me last night and told me with hoarse voice that I should give Rome this revenge for his innocent blood.

[He stabs Aper]

APER.   I'm dying!

CURIUS.   Jupiter on high, what is this?

DIOCLETIAN.   Oh, crown and glory of the world; oh, strong soldiers, companions of my honor, who, fighting by my side, spread your glorious fame from the rose-dawned East to the western sunset, don't be disturbed; for it is not honorable to lose one's composure over the death of a traitor who despoiled our country of the most peaceable king whom our glorious city has seen since the wolf's two nurslings ruled over it. But if I have done wrong and Diocletian angers you, let my best friend draw his sword and break the heart that was capable of this piety.

MARCELLUS.   By the gods, divine courage adorns you, and you deserve for Italy to crown you with the same laurel it gave the fugitive from Troy!

MAXIMIAN.   Soldiers, who in so many victories have seen the sword of Diocletian, mightier and more brave than that of Pyrrhus or Epirus, or that of Alcibiades in great Sparta, honor a man who honors you. Don't give Rome the chance to name a king who will interfere with your martial customs.

DIOCLETIAN.   Soldiers, my rhetoric is not that of Ulysses, nor has my tongue learned flattery. I am your comrade. If your hands crown my temples, you may place the leaves on your own heads. You will all be Caesars. What say you?

EVERYONE.   Give him the laurel wreath!

DIOCLETIAN.   Am I your Caesar?

EVERYONE.   Yes!

DIOCLETIAN.   Then divide the contents of Numerianus and his
  father-in-law's tent — money, arms and jewels — among your-
  selves. I'm satisfied with this clothing and this sword to
  defend you.

CURIUS.   The army does you homage: reign, Diocletian.

EVERYONE.   Hurrah!

DIOCLETIAN.   Let the army march on to Rome.

End of Act One

Act Two

[II.1]

[Fanfare. Enter Diocletian, Lentulus, and a few courtiers]

DIOCLETIAN. I'm very grateful to Rome for acclaiming me as her father and Augustus with such a joyful celebration.

LENTULUS. She is delighted to have you as her emperor. No one recalls when Rome has received a Caesar with so much applause.

DIOCLETIAN. My gratitude will soon show her that she was not mistaken, and also the army, which gave its vote and consent. Today I'll pay the soldiers and distribute a lot of money.

EVERYONE. Long live Diocletian!

DIOCLETIAN. The only thing I want to acquire and store up is goodwill.

LENTULUS *[aside]*. They all get off to a good start, but then they do a thousand misdeeds.

DIOCLETIAN. Let the senate dine with me, and their wives too. Let there be celebrations, as is proper for the people who have acknowledged me as father of the fatherland, Augustus, sacred, heroic, and triumphant.

LENTULUS. What sort of entertainment shall we have, sir?

DIOCLETIAN. Give them whatever they like. Tomorrow we'll have gladiators with all different kinds of weapons; then you can throw slaves and criminals to the beasts. Do you have the beasts on hand?

LENTULUS. Yes, sir.

DIOCLETIAN. What kind of animals do you have?

LENTULUS. A bear, a lion, a tiger, and a serpent.

DIOCLETIAN. That's fine.

LENTULUS. The serpent was found in Libya; Martius brought it in his ship. Rome has never seen it nor any of its kind.

DIOCLETIAN. I think I once saw one like it in Arabia.

LENTULUS. How much like Hercules you are!

[Exit Lentulus. Enter Maximian]

MAXIMIAN. Give me your feet to kiss.

DIOCLETIAN.   You're late, noble Maximian. Knowing my love for you, you shouldn't slight me like this.

MAXIMIAN.   I hesitated to come, oh great one, after seeing you crowned emperor. Your sacred majesty is enthroned so high that we mortals consider you a deity. Hence I hardly dared approach your rising sun for fear that I should be consumed.

DIOCLETIAN.   Maximian, Fortune raises or lowers whomever she wishes. She abandons some, exalts others, and has no constancy at all. She has this immense power over temporal things, but not over our celestial souls, for that is impossible. Thus you should know that, though my status has changed, my soul has not, nor do I think it ever will. Maximian, we two were soldiers and comrades. Together we've suffered thirst, hunger, and hard trials. Now that Fortune has raised me so high, it would be unfair for me to enjoy this bounty without sharing it with you. You may say it's impossible, but it's not, for love equals the lover to the beloved and the beloved to the lover. Since I love you, it's only right that I should share this crown of laurel with you. I name you Caesar. Don't think you can put our friendship behind you. As we were soldiers together, so we shall be emperors. When we shared the bread we snatched from our enemies, we were good friends. Now that heaven's kindness has given us this empire, let's share it as well. I know and trust in God that there will be no envy or jealousy. *[To a servant]* Give me a laurel crown.

MAXIMIAN.   If human words, oh generous Diocletian, sufficed to repay you, I would use them to show the power of your divine worth, but who, sir, could express it rightly? I beg you to consider me a servant in your house, for it is not right that you should try to make me your equal. As your servant, I'll be content.

DIOCLETIAN.   You there! Bring me a laurel crown.

[They bring him a laurel crown]

MAXIMIAN.   Don't think I deserve it, just because I earned your friendship.

DIOCLETIAN.   I place these consecrated leaves on your brow with my own hands, Maximian.

MAXIMIAN.   The fact that you've honored them with your touch is laurel enough for me, for I love you better than the laurel or the Roman Empire, great Diocletian.

DIOCLETIAN.   Sit down by my side.

[As the two men sit down, enter Camilla]

CAMILLA.   The things that heaven ordains in its divine secrets move through such mysterious paths that earth can't understand them. I've followed the Roman army all the way here, selling bread. Back in Asia, Diocletian, who for his good luck and valor is now emperor, was the slave of a Roman. He was so poor that whenever he took some bread, he would laugh and politely promise to pay me when he should be emperor. Since I liked his spirit, I jokingly told him that he would be emperor when he should kill a wild boar. Now he's killed the consul and thus, as a result of a joke, has become Caesar. Since his joke and mine have paid off, I've come to see whether he will repay his old debts. Oh, famous Diocletian, unconquered lord of the world, Emperor of the great Roman Empire! Do you remember the days when, as Aurelius's soldier in Asia, pinched with hunger, you begged me for bread and a thousand times promised to pay me when you reached the position which you now enjoy? The time has come, holy Caesar, since I foretold your honor, for you to carry out as emperor what you promised as a soldier. I am Camilla, I was that peasant and farm girl. If you owe me, pay me now. I have witnesses against you, for Maximian well knows that this is all true.

DIOCLETIAN.   By the supreme deity of sovereign Jupiter, I wish our empire would accept an empress! Meeting you cannot have been without mystery. I'll pay you as emperor, Camilla, for my worth has increased with my change of status. I'll keep my word; tell me the price of your bread.

CAMILLA.   Caesars always give like Caesars.

DIOCLETIAN.   I know that if I were to pay you as you deserve, there would not be treasure enough in Rome. Ask, and let me pay.

CAMILLA.   So you want me to name my price? Very well. I ask that you give me permission to enter and leave your imperial throne room, or wherever you may be, at will.

DIOCLETIAN.   Is that all you want?

CAMILLA.   That is a unique privilege. Give me it, and I'll consider myself paid.

DIOCLETIAN.   Rather, you place me even more in your debt. *[To a servant]* You there! Spread the word that Camilla may approach me at any time, even if I'm engaged in senate business.

SERVANT.   Sir, Genesius is here and begs permission to kiss your
feet.

[Enter Genesius]

GENESIUS.   Your majesty, unvanquished prince, permit me to kiss
your imperial feet.

DIOCLETIAN.   I'm delighted to receive you, for I've heard tell of
your fame.

GENESIUS.   If your glories, your great deeds, your rare, divine wisdom could be expressed in verses and stories, illustrious
Caesar, Genesius would perform your praise, and all the celebrated geniuses not only of Rome but also of Spain and of
Greece would be kept busy writing about them.

DIOCLETIAN.   Today you're to give me an outstanding entertainment. While I dine and the senate honors my table, go and prepare an elegant play.

GENESIUS.   Choose whichever one you like. Would you like
Terence's *Andria?*

DIOCLETIAN.   It's too old.

GENESIUS.   How about Plautus's *Miles Gloriosus?*

DIOCLETIAN.   I'd like a new story that's more original, even if less
artistic. As you can see, I share the Spanish taste in plays. As
long as it's believable, I'm not too picky about whether it follows the rules. In fact I find the precepts too limiting, and it's
been my experience that those who are too careful to observe
artistic dictates never succeed in making a natural plot.

GENESIUS.   I have a play called *The Captive of Love.*

DIOCLETIAN.   It sounds like a generic title. Don't you see that it
would suit all of them, for they all have to have lovers? Who's
the author?

GENESIUS.   Fabritius, a priest of Olympian Jupiter.

DIOCLETIAN.   What's the poetry like?

GENESIUS.   Harsh, priestly, and precious. If he has a chance to call
the sun "eternal lamp," you can bet he won't call it "Phoebus."
He ransacks the odors or spices of both Indies and manages to
find a place for every wild animal and serpent in Libya.

DIOCLETIAN.   They should be his audience.

GENESIUS.   I have a play called *The Quarrel of Marsyas and Apollo.*
The author is Corinthian. He's fantastic at writing wild verses,

but unfortunate in constructing his plots. Still, the parts he gets right are worth hearing.

DIOCLETIAN.   Let's hear what else you've got.

GENESIUS.   I have a play by a Greek who bases all his works on lowering monsters from heaven and raising them back up. The theater has so many drawers and curtains, it looks like a writing desk. There's no chess board like his canvas. If you look at his verses all together, they seem like stones sorted out by a rustic while threshing. But they usually amaze the ignorant mob and bring in more money than good plays, because they talk nonsense; and even if a couple of people take offense, there are still over five hundred who like it.

DIOCLETIAN.   Do you have any tragedies?

GENESIUS.   I have Leonitius's *Electra,* which is better than that of Sophocles. It would make the very stones weep! His tragic verses are graver than those of Seneca. Then I have another by Heraclitus called *Sophonisba.* It's heavenly, just as heroic as Virgil. And I also have *Thisbe* by Cornelius, a great Spanish philosopher and a relative of Lucan's.

CAMILLA.   Don't order a tragedy, may heaven increase your empire from one pole to the other! Since tragedies tell of the fall of empires, it wouldn't be a good omen on your coronation day.

DIOCLETIAN.   Well then, put on a play that you like. I'll leave the choice up to you.

GENESIUS.   I'll perform my own play. That way, if you don't like it, no poet will lose his reputation.

DIOCLETIAN.   I've been told that you're terrific at playing a king, a Spaniard, a Persian, an Arab, a captain, or a consul, but that you're best of all when you play a lover.

GENESIUS.   Acting is just imitation. However, just as a poet can't write convincingly and with feeling about love unless he's in love, since it's love that teaches him the verses he writes, it's the same way with actors. If an actor doesn't feel love's passion, he can't perform it. If he feels the pain of absence, jealousy, insult, the rigor of disdain and other tender feelings of love, he'll play them tenderly, but if he doesn't feel them, he won't know how to play them.

DIOCLETIAN.   Let's go, brave Maximian; we'll honor the senate and let it honor us.

MAXIMIAN.   You, great sir, have already honored Rome with the hopes awakened by your holy rule.

DIOCLETIAN.   Come, Camilla, since you have entrée.

CAMILLA.   If only I had entrée to your heart . . .

DIOCLETIAN.   Do you want my love?

CAMILLA.   I adore your person; I love you more than the greatest treasure.

[II.2]   *[Exeunt all except Genesius]*

GENESIUS.   Love, your burning flame will increase my fame in proportion to your rigor. Your fire, your deep feeling reaches up even to the great emperor, who is already longing to see me imitate what I feel. But it hardly seems proper to use the word imitation for what is truth itself. My will is the play, my intellect the poet of the story I'm inventing, wherein with worthy verse it paints the inevitable stages my thought has gone through. All my mad senses with similar figures have become the actors of my enslaved emotions. My ears play the part of a deaf man, who refuses to listen to reason. My woeful eyes play a blind man, who stumbles through the streets reciting his passion's prayer. My smell is like those people who, according to many writers, live off the fragrance of flowers, since this insane state of mind is nourished by the fragrance of my hope, which is like a flower because it is green and because it is fated to be frustrated rather than bear fruit. My touch plays the part of a madman who tries to touch heaven with his vain thoughts. I ignore good advice and follow that which kills me. In me and with me dwell a thousand asylums full of lunatics, and even so, love thinks them few in number to punish me properly. My taste, which was the greatest and best actor of all, now plays the part of a lover who persists in his mistaken path. And, though it's a comedy of love, unless the playwright does something about it, it won't have a happy ending, for it won't end in a wedding, because suffering makes all the characters tragic.

[Enter Pinabellus]

PINABELLUS.   The company is here now.

GENESIUS.   Did the costumes arrive?

PINABELLUS.   Yes, they're here too.

GENESIUS.   And the musicians?

PINABELLUS.   Florisen said he'd be right along.

GENESIUS.   A fine state of affairs, upon my life! We're always short a musician.

PINABELLUS.   There's still time, for the emperor is eating. Please make sure that they pamper Marcella.

GENESIUS.   Pamper a woman who's driving me crazy and who never loved me? Don't mention that to me, Pinabellus.

PINABELLUS.   Genesius, are you playing a part with me or telling me stories?

GENESIUS.   Would to heaven I were! If I seem to freeze up in the midst of my fire, you must understand that I'm hesitant to love her, for fear that she may altogether despise me. For though jealousy be love, it is a cautious love.

PINABELLUS.   Why don't you just fire Octavius, and then you'll have some peace.

GENESIUS.   You're a little late with that bit of advice. Besides, when I consider all the actors around today, few can come up to him, and he's playing the lead. If I'm so unlucky as to have him leave, I'll only be worse off. If he left the company, Marcella would try to leave too; she'd be sad night and day, she wouldn't study her lines and would skip rehearsals. A person has to enjoy acting. If I upset her that way, once Octavius is gone, she may do something crazy just to make a martyr of herself.

PINABELLUS.   Well then, ask her father for her hand in marriage. I'm sure he'd rather have her marry her boss than some Johnny-come-lately.

GENESIUS.   It's neither honorable nor smart to marry a woman when you know she loves someone else. Marriage is a union of two wills, and there can be no union if one person is unwilling. Caesar has ordered a play about love. I think I'll make it about jealousy, and then it'll suit both of us.

PINABELLUS.   For heaven's sake, don't forget some embraces for Marcella.

GENESIUS.   I've written it specially to give her as many embraces as the chains and ropes with which she's imprisoned my tormented soul. I wrote that mad scene just to give me a chance to abuse Octavius.

PINABELLUS.   Your jealous imagination is unequaled.

GENESIUS.   I've also thrown in her greedy father.

PINABELLUS.   I heard how graspingly he spoke to you.

GENESIUS.   Are they getting dressed?

PINABELLUS.   They're already in costume.

GENESIUS.   Then let my senses get dressed and act for me.

PINABELLUS.   The emperor's coming out. Go inside and get your costume on.

GENESIUS.   Octavius and Marcella are talking to each other. How can I stand it?

PINABELLUS.   They are acting love's desire and anxiousness, while you're acting the comedy of jealousy.

GENESIUS.   Where jealousy is performed, Pinabellus, call it rather tragedy.

[II.3]   *[Exeunt Genesius and Pinabellus. Enter Diocletian, Maximian,*
                                      Lentulus, and Patricius]

LENTULUS.   In the name of the senate and of fortunate Rome, your empire, I congratulate you, and I give you many thanks for the sumptuous banquet.

DIOCLETIAN.   If I am in any way like the good Caesars and hence deserving of gratitude, Rome has already amply repaid me, for she's given me honors that would satisfy the gods. Senators and friends, thank Maximian too for the favors you've received from his heroic hands.

MAXIMIAN.   You alone can do favors for them and for Rome. Diocletian, you exceed everything but sovereign heaven itself in greatness. It is fitting that Rome should speak of you as Jupiter and Augustus, for you have been great enough to divide the empire with me.

DIOCLETIAN.   Let Rome say that I've divided the world with you, if perchance the world is mine to command.

MAXIMIAN.   Your valor is heroic in its greatness.

DIOCLETIAN.   Senators, while I try to forget the harsh exercise of war, I'll give the people entertainment, for I want to win their love.

LENTULUS.   A just desire indeed, though you've already won it, for the people adore you and pray heaven for your long life.

DIOCLETIAN.   I seek only their tranquility and prosperity.

                                                              [Enter Camilla]

CAMILLA.   The play is ready now.

DIOCLETIAN.   Then tell them, Camilla, that our ears are ready to
listen to that image of life. Everyone, sit down and let them
begin.

MAXIMIAN.   Let the singer begin.

PATRICIUS.   This Genesius is a great actor.

        [They sit down. Enter the singer and musicians]   [II.4]

SINGER.   When Diocletian entered Rome,
    Such a great applause did sound,
    We knew our Caesar had come home,
    And with the laurel he was crowned.
    He it is whose mighty deeds
    Caused the army to elect him,
    And now our voices, pipes, and reeds
    Beseech great Mars still to protect him.
    Rome, celebrate this glorious day
    While we the warlike drum do play.
    Princes, of great kings the seed,
    Too oft their heritage do squander;
    'Tis only virtuous word and deed
    Can win them praise both here and yonder.
    Diocletian earned his crown
    By his deeds of heav'nly splendor.
    Now spread abroad his just renown
    And to him your homage render.
    Rome, celebrate this glorious day
    While we the warlike drum do play.

        [Exeunt singer and musicians. Enter Genesius]

GENESIUS.   When Alexander the Great arrived in Athens, he was
eager to meet the great poet Thebanus, for in those days
princes honored poets. Alexander had heard Socrates say that
only Thebanus's pen and tongue could justly praise arms and
letters. Thebanus came to him and removed his gloves at the
door, for it is indiscreet to enter a king's presence wearing
gloves, and anyone who did so would be more of an animal
than a man, like the horse, who is allowed to wear his shoes
before the king only because he can't take them off. The
famous poet carried his gloves in his hands, and when he knelt
before the great conqueror, he was so nervous that he dropped
them. Thebanus heard the noise, and when he looked down
and saw the gloves on the floor, he thought they must be the
king's, so he picked them up, kissed them, and timidly offered

them to the king. Alexander told him: "They are yours," and covered his face so as not to embarrass the poet with his laughter. The poet couldn't say a thing, but just kept bowing. Today, invincible lord — may heaven grant you long and happy years — you have come to Rome as Alexander came to Athens, but in all humility I must say that I was the one who was eager to meet you and not you, me. When I came in with my speech all rehearsed for you, I removed the senses and powers of my soul, as if they were gloves — since the soul fits into them — and put them not in my hands but in my tongue, though indeed the tongue is like the hands, for it is the hands of the soul, since it carries out what the soul disposes and does what the soul commands. When I beheld your divine, august, and sacred presence, sir, all my words fell away from me. I got down on the floor where they were, since it is only fitting that words which fly up to your sun on wings of wax should fall to earth again, and, humbly picking them up, offered them to you. Imitating Alexander's modesty, you said: "They are yours, Genesius. These must be your words, because if they were my praises, they wouldn't be on the ground." I know, unvanquished lord, how low and crude I am, and you must know that if, out of respect for your presence, the three potencies of my soul have fallen from me, I don't deserve blame but rather pardon for humbling them on the ground.

[Exit Genesius]

MAXIMIAN.   That, sir, is Genesius.

DIOCLETIAN.   He's quite an actor! I've never seen such a performance.

LENTULUS.   There's none other like him.

DIOCLETIAN.   Was that all really ad-libbed?

CAMILLA.   Yes, sir, for he's a great poet.

DIOCLETIAN.   That was an admirable comparison.

PATRICIUS.   Brilliant!

DIOCLETIAN.   How well he expressed himself! Camilla!

CAMILLA.   Sir?

DIOCLETIAN.   Give Genesius this ring at once for the prologue, and later he'll get the rest of his reward.

[Enter singers and musicians]

SINGERS.   Lucinda, that your lovely eyes

Aren't the stars up in the skies
Is possible;
But that their soft heav'nly light
Sheds no radiance on the night,
Impossible.
That your sweet celestial mouth
Is not rare coral from the south
Is possible;
But that its fragrance is not better
Than the rose, its color redder,
Impossible.
That you are no sun or Phoebus,
Angel pure, love's unique Venus
Is possible;
But that you have not plucked what's best
From the angels' heav'nly nest,
Impossible.
That the lilies whiter are
Than your beauteous hands by far
Is possible;
But that in them I've not seen
All that's graceful and serene,
Impossible.

[Exeunt singers and musicians. Enter Genesius as a handsome youth
and Marcella as a lovely young lady]

MARCELLA.   Begone, torment me no more.

GENESIUS.   Do I cause you so much pain?

MARCELLA.   So much, Rufinus, that it may be fatal, for my hatred
of you is greater than your love for me.

GENESIUS.   Fabia, if hating me torments you so, imagine how I
must suffer from your hate. There is as great a difference
between my suffering and yours as there is between loving and
forgetting.

MARCELLA.   My pain is much greater.

GENESIUS.   Don't believe it, for you've never been hated in your
life. Everyone loves you, Fabia, and there's just as much difference between being loved and hating as there is between killing and dying. But let's talk about the reason why you're
leaving me.

MARCELLA.   If I were to listen to your complaints, I'd be giving you satisfaction.

GENESIUS.   Fabia, why are you so determined to treat me harshly and reject my love?

MARCELLA.   It's better that you should face the truth.

GENESIUS.   I know very well, Marcella, that the reason why you treat me so unjustly is that you love Octavius. Octavius satisfies you, Octavius pleases you, you ingrate. It is because of him that you're leaving me.

MARCELLA.   Genesius, are you acting?

GENESIUS.   Yes, I'm acting out the pain I suffer from your ill treatment.

MARCELLA.   Why do you call me Marcella, when my name is Fabia?

GENESIUS.   Because I'm really talking to you, because I want your scorn to take pity on my mad love.

MARCELLA.   What should I answer?

GENESIUS.   If you knew how to appreciate my love, you'd answer me better.

MARCELLA.   This is not in the play. Don't you know Caesar is watching us?

GENESIUS.   Don't get angry, Marcella. Cure my suffering!

MARCELLA.   You've got me all confused. Let's go back to where we were.

GENESIUS.   Go back, then, for if my love has its way, today I'll marry you and we'll perform my wedding for Caesar.

MARCELLA.   As for me, I'll act out the hatred that makes me want to leave you.

MAXIMIAN.   Something seems to have upset them. They're just talking to each other.

LENTULUS.   It's probably just that seeing you, sir, has made them forget their lines.

DIOCLETIAN.   On the contrary, I think it's the artifice of this great actor, because being upset is always the best sign that someone's in love.

[Enter Fabritius as an old man]

FABRITIUS.   What greater care could there be in all the world than that of a father who must watch over the honor of his daughter? Where else could Argos be more vividly portrayed than in me, for it's impossible to keep a beautiful woman locked up. Oh, how I could use that fierce-eyed dragon into which Athena changed herself! Fabia, what are you doing here? Since when are you alone in the house with a man?

MARCELLA.   Don't worry, father. This young man was politely asking to see you.

GENESIUS.   May Phoebus guard you, Thebandrus.

FABRITIUS.   What can I do for you, friend?

GENESIUS.   My boldness in love is not new. Your lovely daughter's reputation, which speaks highly of her virtue, aroused this love in me and also made me want to see her and talk to her. I didn't want my father to arrange it. Now I know that I was not deceived, for I've seen the radiance of the sun in her. I implore you, sir . . .

FABRITIUS.   If I didn't remember what it's like to be your age, I'd make you pay in blood for the honor you've stripped from these portals.

GENESIUS.   No, not in blood, for it's not the fault of blood. Rather, I'd repay with my own honor whatever harm has been done by the main offenders – my eyes, my tongue, and my desire. But all these will make good my fault, if there can be any fault where I offer love as my bond, which lengthens the ill but diminishes its pain. Do you find any fault in me? Don't you know that I'm the son of the worthy Consul Patricius?

FABRITIUS.   Since your eyes and tongue so eloquently speak your love, how can I find words to answer you, young man? But since calm is usually more effective in these situations than anger, I'll attempt to respond.

GENESIUS.   Not so fast, begging your pardon! For if you contradict my hope, I'll kill myself right here!

FABRITIUS.   I mean to encourage you, but it's a serious matter to allow a man to be in one's house with one's daughter without first speaking to his father.

GENESIUS.   Then let's go together and tell him of my love.

FABRITIUS.   Let's go, then, for you know my pleasure in this matter.

GENESIUS.   Yes, but be advised that your Fabia is now mine.

FABRITIUS.   As far as I'm concerned, I'll do whatever you like.

GENESIUS.   Then as a token of this joy, let me touch her hand, and
I'll count that as good as your signature on the matter.

FABRITIUS.   Your love is equalled only by your obstinacy. Daughter,
since... What's your name?

GENESIUS.   Rufinus.

FABRITIUS.   ...since Rufinus is determined to be your husband, I
suppose you can let him embrace you, for you're really risking
nothing.

MARCELLA.   What outrageous nonsense!

GENESIUS.   Fabia, may your lovely hand enrich me. Don't let me
die!

[He hugs her]

MARCELLA *[to Fabritius].*   I'm bound to obey you.

GENESIUS.   Even death cannot take the joy of this moment from
me.

[Enter Octavius and Pinabellus]

OCTAVIUS.   Did you see that?

PINABELLUS.   Yes, I did.

OCTAVIUS.   You saw him hug her?

PINABELLUS.   Yes, I saw that too.

GENESIUS *[to Fabritius].*   Now come with me and speak to my father.

FABRITIUS.   We're off then, and may heaven grant it turn out pros-
perously.

GENESIUS.   Is there anything that love can't do?

[Exeunt Genesius and Fabritius]

OCTAVIUS.   Is there anything that jealousy can't do?

MARCELLA.   My Octavius!

OCTAVIUS.   How dare you speak my name, you ingrate! After treat-
ing me like this, you have no right even to look at me, much
less call me by name. As holy Jupiter lives, if Rufinus had tar-
ried here one moment longer, I'd have given him something to
remember his engagement by! Is this what you meant by all
those promises and vain oaths? Is this how all my striving and
my crazy planning ends? Is this what I get for trusting you? Oh,
Fabia, you're just a woman after all, and you're fickleness

itself! Enjoy your new fiancé, and may you wear mourning for him when you've hardly had a chance to enjoy each other! If there's any god in heaven, may he punish the betrayal of earthly lovers that I've heard tonight!

MARCELLA.   Stop, hold your insults, Octavius, for you'll soon see that I'm...

OCTAVIUS.   You dare to call my name?

MARCELLA.   ...innocent. I was distracted, and Rufinus came in the house without any encouragement from me. I gave him my hand out of obedience to my father; but what difference does it make?

OCTAVIUS.   Hold your lying tongue, or you'll soon exhaust my patience and make me kill you.

MARCELLA.   Pinabellus, if a crazy lover won't believe oaths and persuasion, what proof will sway him to cease his madness?

PINABELLUS.   Lovers waste their time in a thousand silly quarrels. If you love Octavius and Octavius loves his Fabia, since you've done him no wrong, he should stop insulting you. Why don't you just figure out how to attain the object of your love, and let bygones be bygones? Love, when once offended, defies reconciliation. There's no other commodity where the middleman is so important. If your father imagined that you really loved Rufinus and therefore obliged you to marry him, the solution is simple. All you have to do is tell your father the truth and break the engagement.

OCTAVIUS.   Any solution you can offer will be risky and vain in our present predicament.

PINABELLUS.   If you trust me, the safest thing to do would be to have me take Fabia away from here. I'll put her on a boat, and you can take her to your land.

MARCELLA.   Octavius, if you doubt for a moment that I'd risk my life and honor for you, I'll know that you don't really love me.

OCTAVIUS.   Then you'll go with Pinabellus wherever he wants to take you?

MARCELLA.   There is no place so distant on earth nor under heaven that I'd not go there with him.

OCTAVIUS.   Then come with him and with me.

MARCELLA.   Forgive me, father, for I'm following my husband.

PINABELLUS *[aside]*.   Today, cruel love, you've been kind to me. Octavius will never see her again, though by stealing her I'm betraying his confidence.

MARCELLA.   Do you promise to marry me?

OCTAVIUS.   I give you my hand and word.

MARCELLA.   Oh, heavens, if only the play were for real!

OCTAVIUS.   Nothing could make me happier.

MARCELLA.   I've lost my place. Oh, I wish we could play this trick on Genesius!

OCTAVIUS.   I admire your faithfulness, Marcella.

MARCELLA.   You'll have better cause to admire it later.

[Exeunt Marcella, Octavius, and Pinabellus]

DIOCLETIAN.   I suspect that those two are not acting.

CAMILLA.   They're just adding their personal interpretation to the play.

LENTULUS.   Now the servant intends to betray his master. Octavius, who thought he was betraying her father, is himself betrayed. So now both Rufinus and Octavius will be bereft of Fabia, and Octavius's slave will enjoy her.

PATRICIUS.   Here's the father again.

[Enter Genesius and Fabritius]

FABRITIUS.   I'm glad your father is so happy for our families to be joined.

GENESIUS.   He'd be no father but an enemy if he were not honored to have you in the family.

FABRITIUS.   Since he's indisposed, he wants us to bring my Fabia to him.

GENESIUS.   Now even sickness wouldn't bother me.

FABRITIUS.   The health of lovers consists in getting what they want. We've already discussed the dowry, and I see that he's more interested in my daughter's virtue.

GENESIUS.   He wants only what I want.

FABRITIUS.   I can see that you care only about your love.

GENESIUS.   The beauty that is consuming me in its flames is all the dowry I could ever want.

FABRITIUS.   Noble birth, virtue, and beauty are no mean dowry.

[Enter Celius]

CELIUS.   Oh, wise Thebandrus, if Rome has ever held up your prudence and your rare intellect as an example for others, she'll
know now whether she was right or wrong in saying that you
exceeded the great Cato. Thebandrus, your daughter Fabia,
seeing that you were marrying her to this young man and
driven to despair by love for another, has taken the worst possible course of action – but when did lovers ever choose wisely?
She's run off with Octavius – that same Octavius, more noble
than clear-headed, whose haughtiness and pride once angered
you right in this very street. They're on their way to the port of
Ostia, where I've heard they plan to embark on a ship . . .

FABRITIUS.   Stop! Don't go on, Celius.

GENESIUS.   Celius, what are you saying? Fabia ran away with
Octavius?

CELIUS.   They've probably already set sail, Rufinus.

GENESIUS.   Then how am I still sane? Go, Thebandrus, for God's
sake, go and stop them from sailing, or else let me kill myself
before they do. Could anything worse happen?

FABRITIUS.   Rufinus, killing yourself won't make me feel any better. The only thing to do is to go after Fabia. Octavius is stealing my honor.

[Exit Fabritius]

GENESIUS.   Celius, stay a moment; stay and tell me who told you
about this terrible event? Who saw it? Who told you?

CELIUS.   I did, I myself, for I've come just now from the port, where
I saw the two of them arranging to take a boat to Spain.

GENESIUS.   Stop her, enemy heavens! Oh, foamy sea, raise the
sands of your depths up to the stars that adorn the gilded firmament! Rough winds, toss the waves so strongly that the
oppressed sea will bellow and the elements will tremble. Let
the skilled navigator, who is always watching his needle, lose
his bearings. Let the ship go out of control! Let the pilot be so
upset that he won't listen to the helmsman nor will the crew
obey him! Let the sails fall, broken by the raging winds! Let the
ropes and parrels, topsails and cables, break asunder! If you go
out on the poop deck with your haughty lover, as Helen did
with Paris, flinging insults at the Greek, let the wind strike
such a blow that he will be torn from your arms to the bow-
sprits of the prow, flying right over the cross-jack sails! But

stop her, heavens, for my thoughts will hardly reach her while she's on the water and I am on fire!

LENTULUS.   Rein in your love, Rufinus, for it's like a runaway horse!

DIOCLETIAN.   He's a good actor.

MAXIMIAN.   Extremely good.

DIOCLETIAN.   I've heard Genesius praised for this kind of mad scene, with its anguish and lovesickness.

LENTULUS.   You see this play?

DIOCLETIAN.   Yes indeed.

LENTULUS.   By holy Mars, everything you've seen — even the way Genesius can move the very stones when he plays a lover — is nothing compared to what he does in the role of a Christian, one of those followers of Christ, who, leaving behind our gods, sacrifices, and sacred fires, get themselves baptized. He plays that part so boldly, with such forceful words, that he'd just amaze you.

DIOCLETIAN.   Tomorrow, just to ridicule those who deny their rightful incense to Mars, Venus, Jupiter, and Mercury, I want Genesius to put on a play about them and play one of them. I'd like to see in person a Christian who stands firm amidst so many tortures.

MAXIMIAN.   Listen, he's going back to his harangue.

GENESIUS.   Holy Neptune, since you are the god of the damp element and Proteus obeys you, raise your head out of your cold urns, on whose icy walls algae serve as tapestries and mother-of-pearl as chairs, oh raise that head crowned with pearls and young coral, and hold aloft your trident. Stab all your salty empire so that, all enraged, it may dash that ship on the jagged reefs, on the Euripos and Scyllas, like one who breaks a mirror. Celius, what shall I tell you, Celius? Tell them to come out, for I've ad-libbed this whole bit. This is the end of my speech.

CELIUS.   I can see that you're talking without a script.

GENESIUS.   They must be getting dressed. I'll repeat that last line again: "But stop her, heavens, for my thoughts will hardly reach her while she's on the water and I am on fire!"

[Enter Fabritius]

FABRITIUS. Why are you wearing yourself out with shouting, Genesius?

GENESIUS. Is that all you have to say when I'm performing before the greatest emperor in the world?

FABRITIUS. I'm trying to tell you that Octavius has really done this wrong and has dishonored me in fact.

GENESIUS. What?

FABRITIUS. He's carried off Marcella.

GENESIUS. Take off that beard and announce that this is the end of the play, and then start watching your tongue.

FABRITIUS *[to Diocletian]*. Punish them, invincible lord, for Fabia – that is, my daughter Marcella, whom Genesius loved – has done the very thing they were pretending to do. In very fact they've left the palace, and in the short time it took for this first scene, they've disappeared, and no one knows where they've gone.

GENESIUS. If, as Rome believes, Caesars are obliged to do justice, order them to go after the traitor who carried off this woman. If you will not, we cannot end the play.

DIOCLETIAN. Is this an act and part of the play or do you want to show us, Genesius, that with this sort of trick you can make us actors too?

GENESIUS. No, sir, it's quite true that Octavius loved Marcella, and since her father preferred me, because I'm his manager, they drew this plan, so that I myself have written the trick they've played on me.

DIOCLETIAN *[stands up angrily]*. By Jupiter, I suspect that you want me to perform, and I don't know whether to refuse. Are you acting or not?

[Enter Pinabellus]

PINABELLUS. Sir, Octavius has come back. Your majesty may be seated.

FABRITIUS. Now what do you think, sir? Isn't Genesius a terrific actor?

GENESIUS. From now on, great sir, I'll be your strongest supporter, since you've helped me so well to get what I wanted.

DIOCLETIAN. I'm happy with the trick, and since I've played my part so well in your play, there's no need for the treasurer to pay you.

GENESIUS.   I'll consider it payment enough to have had you as my fellow actor.

DIOCLETIAN.   You can stop the play here, but come back tomorrow so that I can show you favor, since today I've been an actor too. I want to see how you play a Christian.

GENESIUS.   Sovereign Lord, you'll see me at my best.

DIOCLETIAN.   Let's go, Maximian.

MAXIMIAN.   I think you enjoyed Genesius's wit.

DIOCLETIAN.   He's sharp as a tack!

[II.5]   *[Exeunt Diocletian, Maximian, Lentulus, Patricius, and Camilla]*

GENESIUS.   Tell me, Pinabellus: is it true that Marcella came back, or did you just say that so as not to offend the emperor?

PINABELLUS.   I'm sorry to take the wind out of your sails, but even the emperor wouldn't be able to catch Octavius now.

GENESIUS.   Then my misfortune is true.

PINABELLUS.   Yes, and it's true that she and Octavius are setting sail.

GENESIUS.   Oh, what terrible confusion!

PINABELLUS.   Be careful; he may still hear you.

GENESIUS.   Now I can say again, since my suffering is real: "But stop her, heavens, for my thoughts will hardly reach her while she is on the water and I am on fire!"

End of Act Two

Act Three

[Enter Diocletian and Camilla]   [III.1]

CAMILLA.   As soon as I saw that my ruse had worked, I told you of my feelings.

DIOCLETIAN.   I had always kept my love a secret.

CAMILLA.   When I asked you for entrée to the palace, it was only in the hope that my love might find an entrée to your heart, as it now has.

DIOCLETIAN.   Camilla, you sowed the wheat of your bread on fertile soil, even though it was in wartime.

CAMILLA.   I sowed my hopes with you, and I've reaped such a harvest as reward for my zeal that I could offer heaven the whole earth in tribute. Whatever power I have is derived from your sacred majesty, Caesar, for you are the master of the world, but I am mistress of your will.

DIOCLETIAN.   Camilla, I was so happy to learn that all you wanted was to be able to enter my rooms so that you could see me, I was so impressed by your great heart and by your scorn for money, when my whole treasury could hardly have repaid you adequately for your bread, that I considered you a wife worthy of a Caesar and gave you entrée to my soul.

[Enter Rutilius]

RUTILIUS.   Now you can come out and see the wild beasts they've brought you, while preparations are being finished for the spectacle in Rome's great amphitheater.

DIOCLETIAN.   Are the beasts very strange?

RUTILIUS.   Rome has never seen any like them. Shall I tell you about them?

DIOCLETIAN.   I'd like to hear it.

RUTILIUS.   Daring leontocomes, as those men who tame lions are called, have brought back two black ones, a species found only in Syria. The ones we have in Italy are always red, with brown and blood-colored spots. They've brought two bears from Mysia — females, because they're fiercer — the biggest ones that Rome has ever seen. There's a famous wild boar, from whose tusks the snakes flee. He's so big that you'd think he was the one who killed Venus's lover Adonis. He was born in Macedo-

nia, where the fiercest ones come from. There's an Indian cir-
copithecus, which has the hair and beard of a man and a white
face, but the rest of his body is black. He climbs up in pine and
yew trees and mocks those who pass by with laughter and wild
gestures. They've brought a cynoprosopus, with the head of a
dog and the body of a man. He's extremely light-footed, has
keen eyesight, and is altogether, from head to toe, just as Virgil
described him in his elegant verses. There's a camelopardal,
which the black Ethiopians call "nabim," and which Rome used
to have in the circus games when Caesar was dictator. From
Germany they've brought a bison whose head has only one
horn, and two panthers like those that Scaurus brought Pom-
pey, with their hide as varied in color as the tail of a peacock.
They've brought a strong tiger from Persia, like those that the
Emperor Claudius once brought to Rome; his body is as large
as that of two big lions, and he has three rows of sharp teeth
that could bite an ox in half. There's a tarantus, whose skin has
the colors of all the different trees and whose horns have a
thousand branches. The Scythians make shields out of his
hide, and once they're dried and stretched on wood, no iron can
penetrate them. There's a Pegasus the size of a Friesian horse,
or one of ours, whose tail is so beautiful that it brings a high
price. The Indian women wear it as a wig. There's a pathaga
which in its shells, back, and skin resembles the Egyptian croc-
odile, that weeps before it kills. There's also an onocentaur,
with a man's face and the body of an animal, and a monopus
from the mountains of Africa that's as tall as a camel, and a
hermaphrodite hyena, which fools shepherds by imitating their
own voices. They've brought the fearsome catobletus, a horri-
ble sight to see, and the swift African monkey, with hands and
feet like a man's, a horned rhinoceros, who looks like a cliff
from a distance, and a dragon who kills an elephant by holding
on to its chest, but doesn't get to brag about it, because the
dead elephant falls on him and kills him. Servilius has brought
these and others that I haven't mentioned from distant realms
for your pleasure. Even talking about them scares me!

CAMILLA.   Nevertheless, I know another animal that's fiercer than
any you've mentioned.

DIOCLETIAN.   Fiercer than all of those?

CAMILLA.   Greater, more cruel and invincible.

DIOCLETIAN.   What's its name?

CAMILLA.   Love.

DIOCLETIAN.   You're right, for it is terrible.

CAMILLA.   If they didn't bring that animal, none of the others could
be so strong, for it fears not even death when it seeks its pleasure. Those other animals can harm our bodies, but love
wounds the soul, which is a stranger and more stupendous cruelty.

DIOCLETIAN.   Go, Rutilius, and say that I'm coming to see the animals.

CAMILLA.   Well, sir, I don't feel like going with you if they're going
to throw captives to those beasts with the exotic names.

DIOCLETIAN.   Why not?

CAMILLA.   I don't want to see them killed, because they are men.
Since you are a man, for your sake I respect all men.

DIOCLETIAN.   Thanks for the compliment.

CAMILLA.   That's what you mean to me. I know, Caesar, that you're
counted as a god, but that will be when you rest in peace.

DIOCLETIAN.   Enough said about this beastly celebration, for cruelty is not amusing. But let Rome see them for the novelty of it.

CAMILLA.   Only you could be so discreet and charming.

DIOCLETIAN.   You, there! Summon the actors.

RUTILIUS.   I think they've already been ready for an hour and a
half.

[Exit Rutilius. Enter Genesius]

GENESIUS.   If your majesty wishes, we'll begin right away.

DIOCLETIAN.   Oh, Genesius! We haven't seen you since Marcella
treated you so cruelly. What happened to them?

GENESIUS.   Her father went after them and decided that the best
punishment would be to marry them.

DIOCLETIAN.   If they really loved each other, you're absolutely
right, Genesius. Why is it that pleasure vanishes when lovers
are finally together? Love lasts as long as the fear of losing
what one loves; for as soon as it's possessed, love wanes. What
have you done with them?

GENESIUS.   I took them back, married them, and forgave them.

DIOCLETIAN.   But doesn't it make you jealous to see them?

GENESIUS.   Yes, but my jealousy goes away when I consider that
she's his wife.

DIOCLETIAN.   You managed to forgive them? Well, you behaved
just like a poet, for poets know more than just loving.

GENESIUS.   We poets are obliged to pardon love's frailties.

DIOCLETIAN.   You're right, since your souls are more affected by
your cares because of the sensitivity induced by writing poetry,
and because you can better imagine how love affects a man's
soul. Do you know what play I want you to perform?

GENESIUS.   I await your pleasure.

DIOCLETIAN.   The story of a baptized Christian, because I've heard
it's one of your best roles.

GENESIUS.   As you like, sir.

DIOCLETIAN.   Then set the stage while the senators are coming and
get everything ready. And you, my love, come to the garden,
whose flowers I see in your cheeks, and while we're waiting,
you can enjoy the fountains.

CAMILLA.   You are my only joy and joy enough for me.

[III.2]   *[Exeunt Diocletian and Camilla]*
GENESIUS.   True love my life such bitter torments gave,
Spoiling the blooming springtime of my days;
Lost on its sea of lies as in a maze,
I've come not to the harbor but the grave.
    And though this fire still burns among the coal,
With each new day I feel its pain the less.
I loved, was jealous, but now 'twas foolishness,
I know, not love, that near destroyed my soul.
    Lovers, while still deceived, may well attest
A faithfulness whose substance is illusion,
But the rejected lover knows no rest.
    And what if he still loves, in his confusion?
His jealousy is of love too great a test
And soon will strip him of his last delusion.

[Enter Marcella]
MARCELLA.   Tell me, boss, what play are we going to perform?

GENESIUS.   The play of your false love.

MARCELLA.   It can't be false, since it's the greatest love in the
world.

GENESIUS.   Was your love for me false?

MARCELLA.   Yes, it was a passing infatuation, but the love I feel now is the real thing, for I'm faithful and true to the man I love.

GENESIUS.   Indeed. Then I suppose it was only proper that you should have attained that love by trickery.

MARCELLA.   If it were my fault, I'd feel sorry for you and try to make up for the harm I've done. But you're the one who wrote the play in which my character, Fabia, loved Octavius. You showed me the way, so it's your fault.

GENESIUS.   I wrote that you ran away with Octavius, whom you deeply loved, so that I could better feel the pain of your cruelty to me, not so that you would really run away.

MARCELLA.   Well, I saw it differently.

GENESIUS.   I should like to make you happy . . .

MARCELLA.   Love knows that I am.

GENESIUS.   . . . and not to be sorry.

MARCELLA.   How could I be sorry?

GENESIUS.   I'm frightened of your fickleness.

MARCELLA.   If I change my mind, then I'll love you.

GENESIUS.   And will you change your mind?

MARCELLA.   I don't know.

GENESIUS.   Well, at least I have hope then.

MARCELLA.   Aren't you sure I'm fickle?

GENESIUS.   Yes.

MARCELLA.   Well, then, it shouldn't bother you too much to wait a while.

GENESIUS.   I've taken your words to heart. I think I'll use them as the basis for a play, for it sounds as if we're both acting. *[Aside]* By God, she drives me crazy!

[Enter Octavius]

OCTAVIUS.   They're talking to each other!

GENESIUS.   Your husband!

MARCELLA.   So what!

OCTAVIUS.   I wonder what play they're planning now. Honor and shame, restrain my jealousy!

dwells in glory is to adore him and receive the holy water of baptism, . . .

FABIUS *[Aside].*   He's distracted; he must not have noticed me.

GENESIUS.   . . . why should I be surprised that his name penetrated my senses so powerfully?

FABIUS.   Oh, sir! It's terrible to have her lord it over us. *[Aside]* He doesn't realize I'm talking to him.

GENESIUS.   And they say that there's a hell for those who refuse to follow him! Then it's no wonder that a Christian would die for Christ!

FABIUS.   Oh, sir! Don't waste your time like this. The emperor is coming. We're going to need you, for Marcella told me to learn the angel's lines. The only thing she cares about is Octavius.

GENESIUS.   Yes, God did speak to me and showed me who He is!

FABIUS.   What God? I'm the one who's been talking to you about this angel business.

GENESIUS.   Oh, Christ! You spoke to me through an angel!

FABIUS.   I don't know any Christ, and I've never seen him.

GENESIUS.   Is that you, Fabius?

FABIUS.   Yes, sir.

GENESIUS.   Forgive me. I got carried away playing the Christian and lost my senses, thinking that an angel was whispering in my ear.

FABIUS.   What angel? Are you all right? I'm the one who was talking to you about the angel.

GENESIUS.   You were talking about an angel?

FABIUS.   Yes, Genesius.

GENESIUS.   Then I must have been mistaken about the voice, for I thought it came from heaven.

FABIUS.   Since Marcella is your heaven, and she was supposed to play the part of the angel, I suspect that while thinking about her, you thought you could find heaven on earth. Well, listen to this. She doesn't know the angel's lines, and she's ordered me to learn them.

GENESIUS.   This is a serious matter! Either heaven is playing tricks on me or I heard its sweet voice. But it must have been Fabius, who was talking to me about the angel. So Marcella can't play the part?

FABIUS.   She says not, because she didn't go over it yesterday.

GENESIUS.   And can't you do it, Fabius?

FABIUS.   No, by God! It's been a year and more since I last played that part.

GENESIUS.   Here comes Caesar. Since there's no alternative, you'll just have to do the best you can. Come here and go over it with me.

FABIUS.   I know we're going to make a mess of it.

GENESIUS.   Oh, Christ, since you are God, you'll lead me to you, for I shall follow you from now on!

[Exeunt Genesius and Fabius. Enter Diocletian, Camilla, Maximian,   [III.4]
Lentulus, and Sulpicius]

DIOCLETIAN.   Let Camilla sit between the two Caesars.

CAMILLA.   Just see what fickle Fortune can do! She raises some up and casts others to the ground. Where will this spinning end? Sometimes she twirls at the end of a rope, and sometimes she rises to the very sphere of the wind.

MAXIMIAN.   Where will it all end, Camilla? Just in being contrary.

LENTULUS.   He's right, for yesterday she was selling bread to the Roman army and Diocletian was even worse off, and today she sits between two emperors to hear a play.

MAXIMIAN.   Sit down. People are already coming in.

CAMILLA.   Be quiet, the tragedy is beginning.

[Enter singers and musicians]   [III.5]

SINGERS.   Christ descended to this world
From the bosom of the Father,
And the Word took on man's flesh
In the holy Virgin Mother.
In his blood he wrote the law,
Leaving us his holy Gospel,
Righteous guide for Christian souls
Who his blessèd faith have followed.
Their beliefs have strengthened them
Atrocious martyrdom to suffer,
And through death reborn to heaven,
They in bliss behold his kingdom.

[Exeunt singers and musicians. Enter Marcella]

MARCELLA. Naturalists have written such strange things about the noble elephant that they seem incredible. They tell of one who learned to write, which is amazing and wonderful, since there are many men who have never learned to form letters. They say that he wrote in the sand: "I'm the one who wrote these letters, which may serve as my epitaph and recall my invincible achievement for posterity." They tell about another elephant who was so ashamed because another had shown himself braver than he in battle that he threw himself off a high cliff, where the dolphins wept over his sandy grave. But these are just isolated instances. More generally, two things are said of them which serve my purpose. First, they say that when they have to pass through a flock of simple lambs, they move them out of the way with their trunks so that no one will step on them. Second, when they come to a river, they don't allow the big and small ones to cross over at the same time, for fear that the water level will rise with their bulk and drown the little ones. Hence, they have the little ones cross first, and none of the full-grown ones enters the water till they see that the little ones have reached the other shore safely. Now I see two Caesars before me who have come to this place where, like little lambs, Genesius and his poor and humble flock, who humbly serve you, are passing through. It would be well for you to move us out of the way with your invincible hands, since the whole world owes you homage. And if we must cross so great a sea—no little Po or Tiber—it's up to you to make sure we don't drown. Hence, it is only fitting that we ask your majesty to draw aside and watch us till we reach the other shore in safety, so that none of us may be endangered. Do us this favor, please, for it would be wrong for illustrious men to refuse a reasonable request, especially when it comes from a woman.

[Exit Marcella. Enter Genesius, being dragged away by a captain and three soldiers]

GENESIUS. If you must arrest me, you could at least treat me with respect.

CAPTAIN. You talk too much, Leo.

GENESIUS. I don't mind dying. And although, for Christ's sake, I'm not ashamed of going to jail, I spoke out of human pride and now I'm sorry. Mistreat me, despise me, show me your rage, insult me and injure me, and drag me away as you please. For the sake of Christ, it all seems glorious.

CAPTAIN.   Let go of him, for such humility deserves some pity.

DIOCLETIAN.   How well the story begins! This Christian is going to jail.

MAXIMIAN.   Genesius is playing the part, and you'd think he was a Christian himself and all of this was really happening.

GENESIUS.   Oh, Lord! I already belong to you, but if only I were baptized, in case perchance they fail to martyr me! But if not, I know that the baptism of blood will suffice.

SOLDIER.   That's not in the play.

CAPTAIN.   You never know what he's going to come up with.

SOLDIER.   He's certainly in good form to play before Caesar.

CAPTAIN.   You're right, but they say that a chance like this always sharpens the wits.

[An angel appears on the balcony]

ANGEL.   God has heard your thought, for God understands the language of the heart, Genesius, and He is pleased with your desire. Come up, come up to me. I want to baptize you.

GENESIUS.   Lord, though I don't know how to talk, I'm sure you'll understand me, since you understand the silent language of my thoughts. Take me where you wish.

[He goes up to the balcony]

CAPTAIN.   I don't know how this scene is going to end. This is not the way we rehearsed it.

SOLDIER.   He just keeps improvising without warning us.

CAPTAIN.   Where's he going?

SOLDIER.   I don't know, but now he's gone behind a curtain.

DIOCLETIAN.   Now Genesius is pretending that after he accepted Christ, who is the God of the Christians, that angel came to him, to teach and defend him.

MAXIMIAN.   What a lot of rubbish!

DIOCLETIAN.   But that's just how they are.

CAMILLA.   He must be praying now, for they offer prayer instead of sacrifices. I know, because I went to see them once. They were all watching one who was lifting up a host, because they think their God comes down in that form.

DIOCLETIAN. I don't understand it.

[Music plays, and a blue curtain opens on the balcony, revealing an angel kneeling, holding a bowl. Another angel is holding a pitcher aloft, as if he'd just finished pouring the water. Another angel holds a lighted candle, and another, a hood]

GENESIUS. Divine Lord, you who, as God, can see and hear our thoughts; who can make a prophet of an ignorant Amos, and who can raise a Lazarus from the dead; you who could raise a thief to heaven because he converted with his dying breath; oh, second person of the Trinity; you who saved Jonah from the sea and revealed yourself at Emmaus: bless this bread, for it is yours. Perform with me from now on. You play the mercy of Jesus, and I will play the martyrdom of Genesius.

[The curtain closes]

DIOCLETIAN. That scene was outstanding.

MAXIMIAN. I liked the special effects.

CAMILLA. His acting is marvelous!

LENTULUS. You can't tell the difference between this and the real thing.

CAMILLA. He was just like a Christian in the baptism scene – so humble and with his hands joined in prayer!

DIOCLETIAN. You'd think he was really one of them.

GENESIUS *[as he comes down from the balcony]*. Lord, now that you've done me so great a favor, gird my breast with your armor of love, for if I'm strengthened by you, I know that death, which will come dressed in my frailty, cannot defeat me. Come, friends, for I'm ready to go to martyrdom joyously.

CAPTAIN. What in Apollo's name is going on? This line and this whole scene are nowhere in the script.

GENESIUS. What can I say, except that I'm going to die, which is a line that will keep me in line. God put these lines in my part. I couldn't follow Him if I skipped these lines. This way I'll follow Him in the play up to His banquet table and the heavenly life which I hope for in Him. We are all actors, and any actor who leaves out these lines will surely be raked over the coals.

CAPTAIN. Tell the prompter to give Genesius his line, for he's completely lost.

SOLDIER. Hey, there! Give us the lines!

GENESIUS.   Can't you see that heaven is already prompting me? As
     soon as I heard an angel say from behind a blue curtain, "Come
     up, Genesius, come up to me," I realized that the script was all
     wrong. Where it said "God," it should have said "the devil";
     where it said "grace," it should have said "sin"; and where it said
     "heaven," it should have said "hell." If I had followed the script,
     that's where I'd have gone. The script said "bliss," where it
     should have said "eternal weeping." But once the angel from
     the heavenly dressing room prompted me and taught me what
     I needed to know, I spoke my lines for God, not forgetting the
     "Hail, Mary," which was also in my part. The audience on high
     knew that I played the role with all my heart, and I won them
     over so much that now they're taking me to heaven. From now
     on I belong only to God, and, since I've performed his faith,
     heaven has decreed that henceforth I shall be known as the
     supreme actor.

CAPTAIN.   Prompt him, for he's lost. Everything he's saying is ad-
     libbed.

GENESIUS.   Yes, for I've learned all of this miraculously.

                         [Enter Fabius, dressed as an angel]
FABIUS.   Genesius, God has sent me to speak to you.

CAPTAIN.   That scene's already over, and we can't go back to it. We
     already did the baptism bit.

FABIUS.   What are you talking about? I haven't even been on stage
     yet.

CAPTAIN.   You have so!

FABIUS.   Who, me?

CAPTAIN.   Yes, you.

FABIUS.   Me? What are you talking about?

CAPTAIN.   Yes, it was you.

DIOCLETIAN.   Actors, have you forgotten that I'm here?

GENESIUS.   It must have been my fault, so there's no reason for you
     to get upset at them.

DIOCLETIAN.   If you don't know the play, why are you putting it on?
     And why are you arguing with each other in front of me?

CAPTAIN.   Didn't your majesty see the angel?

DIOCLETIAN.   Yes.

CAPTAIN. Well, he insists that he hasn't come on yet, and he wants to go back to that scene.

DIOCLETIAN. I can see that.

FABIUS. Great Caesar, if anyone can prove that I've already come on, you may cut off my head!

DIOCLETIAN. But I saw you myself!

CAMILLA. What are you talking about, man? I saw you too; we all saw you.

FABIUS. Gentlemen, it wasn't I. Listen to me, it wasn't I.

MAXIMIAN. Be quiet, you fool. You must be mad.

GENESIUS. He's right. It was a heavenly being with a divine voice that played his part.

DIOCLETIAN. What do you mean, a heavenly being?

GENESIUS. An angel who showed me a heavenly book, wherein I saw all that I've learned, which is the part I've just been playing. Caesars, I am a Christian. I've already received holy baptism. This is the role I'm playing, for the playwright is Jesus Christ. Your wrath is written in the second act, and when the third act comes, I'll play my martyrdom.

DIOCLETIAN. Is this for real, Genesius?

MAXIMIAN. Tell us, Genesius: have you lost your mind?

GENESIUS. I mean what I am saying, tyrants!

MAXIMIAN. Kill him!

DIOCLETIAN. You impudent dog! Do you deny Jupiter?

GENESIUS. Yes, for he's lewd and unworthy to be called a god.

DIOCLETIAN. Then I am ready to say my speech. You will die in a play, since you've lived in a play. I'll sit now in judgment. Bring him here.

GENESIUS. Bravo! I do deny your gods and I adore Christ, the living God.

DIOCLETIAN. Then I sentence you to death. That was a short trial! My part is done. Let Lentulus and Sulpicius arrest and question all your company. Come, Camilla, let's get out of here.

MAXIMIAN. You wretch, why have you forfeited Caesar's grace?

GENESIUS. I have God's grace.

[Exeunt Diocletian, Maximian, and Camilla]   [III.6]

LENTULUS. What have you said and done?

GENESIUS. That I adore Christ, that I'm a Christian, that I follow his faith.

LENTULUS. Guards, come here!

[Enter two guards]

GUARD. Sir!

LENTULUS. Tie Genesius up tight and take him to jail.

GENESIUS. Today, oh Jesus, I bless your holy name, for you have granted my wish.

[They take him away]

LENTULUS. Summon the actors, and have them come in one by one. Make sure that none escapes.

SULPICIUS. Have you ever seen such nerve? To play a trick like this on Caesar!

LENTULUS. Only a Christian could be so shameless.

[Enter Marcella]

MARCELLA. What do you want of me?

LENTULUS. State your name.

MARCELLA. Marcella.

LENTULUS. What was your job in Genesius's company?

MARCELLA. Didn't you see? I played the leading ladies.

[Enter Octavius]

LENTULUS. Who are you?

OCTAVIUS. Her husband.

LENTULUS. What parts do you play?

OCTAVIUS. Leading men.

[Enter Sergestus]

LENTULUS. And what do you do?

SERGESTUS. I play knaves, poor little soldiers, braggadocious captains, and the like. I play most anything they need.

[Enter Fabius]

LENTULUS. Any you?

FABIUS. I play boys, princes, and other parts that call for a young man.

LENTULUS.   Such nice answers! This is some way to get to the root of why they offended the emperor! What do you do?

[Enter Albinus]

ALBINUS.   I play the comics, but I'll be tragic, not funny, if you take out your anger on me. I also play shepherds when a lady gets lost in the mountains and cries out for me.

[Enter Salustius]

LENTULUS.   What parts do you play?

SALUSTIUS.   I play the traitors.

LENTULUS.   I wouldn't want to meet up with you in a dark alley!

SALUSTIUS.   Sir, I'm of a good family. I've never really betrayed anyone.

[Enter Fabritius]

LENTULUS.   And you, my good man, what parts do you play?

FABRITIUS.   I play the fathers and kings, serious roles.

LENTULUS.   Then play my part.

FABRITIUS.   Sir, I don't know the laws.

[Enter Celia]

LENTULUS.   What parts do you play?

CELIA.   I'm the supporting actress. I play maids, shepherdesses, and Moorish women.

[Enter the property man]

LENTULUS.   Who are you? What's your name?

PROPERTY MAN.   I'm the property man, and my name is Ribete.

[Enter Martius]

LENTULUS.   You're the last one. What about you?

MARTIUS.   I'm the gravedigger.

LENTULUS.   What's that?

MARTIUS.   The one who carries off the dead.

LENTULUS.   It pains me to see you. It would be cruel to arrest you. Give me a straight answer. Are you Christians?

EVERYONE.   No, sir.

LENTULUS.   In that case, I sentence you only to be exiled from Rome.

MARCELLA.   I prostrate myself at Caesar's feet and humbly beg his pardon.

LENTULUS. Leave at once.

OCTAVIUS. We won't delay a moment more in Rome, sir.

LENTULUS. I'll tell the emperor you're leaving.

EVERYONE. We'll go together.

[Exeunt omnes. Enter Genesius in chains] [III.7]

GENESIUS.   My God, I only meant to play the role
  Of Christian, but You took me at my word.
  I thought 'twas all a game, but now I've heard
  The stakes were high; I risked my very soul.
    How could I know that heav'n's exalted throng,
  Assembled in the theater on high,
  Attentively beheld the play as I
  My moment of applause sought to prolong?
    But You took pity on my heartfelt zeal
  And saw how I outstripped the other players.
  I played my part so well that it seemed real.
    "Bravo!" You cried, "since this Genesius dares
  To play a saint, we'll hark to his appeal;
  He'll have the part, we'll answer all his prayers."

[Enter Sulpicius and a warden]

SULPICIUS.   These are Caesar's orders.

WARDEN.   Then take him away. What are you waiting for?

SULPICIUS.   When he comes back from the circus, he wants to see him impaled.

WARDEN.   Genesius, you must have put on a terrible show for the emperor!

GENESIUS.   Since I've changed masters, I perform God's truth, for it's foolish to fear and respect a mere man.

WARDEN.   I can't believe that someone who used to make so much fun of the vain martyrdom suffered by the Christians would insist on being a Christian himself!

GENESIUS.   I belonged to the company of the devil. He's a proud but cruel actor, but he made a big mistake when he tried to play God, for those two roles are far apart. Now I've joined the company of Jesus, God the Father, and the Virgin Mary. The Holy Spirit guides me to the two Persons from whom He proceeds. I'm now with John the Baptist, who can play shepherds in the wilderness and make such harmonious music that it exceeds that of the spheres. In this company is John, who speaks so eloquently; and the great poet David, who wrote a perfect play of canticles. Peter plays the part of supreme pontiff, and Saint Bartholomew, that of a man who was flayed alive. Although Magdalen wandered from the strait path, she returned when God gave her the cue. There's the famous thief Dimas. His part is very short, but he said more in it than Solomon in all his books. There's a brave Samson. Christopher will play the giants, and Ildefonsus will play Mary's wardrobe master, with stars for diamonds. Gabriel plays messengers to Mary — lucky Gabriel, to play opposite God's spouses! Paul plays brave and fierce men who are soon disarmed. Francis plays those who imitate God, and Nicodemus is the gravedigger, but the dead he buries rise again. In my former company Judas played the traitors, and Roman emperors played cruel and tyrannical rulers. Lucifer played liars and stubborn men. The world knows how to costume a romantic lead, and the flesh how to perform amorous women. The sinner buries the dead, but they don't return to life.

SULPICIUS.   I can't tarry any longer. You can say all this later, for the emperor wants to see you when he comes back to the palace.

GENESIUS.   God wants to see me, for He chose to make me His in order to teach the devil a lesson. He wants me to act and sing, and by dying, I'll go on acting in heaven, for Genesius will henceforth be known as the supreme actor.

[III.8]   *[Exeunt omnes. Enter the actors who are leaving Rome, some with packs on their shoulders and others with costumes and props]*

OCTAVIUS.   Rome, farewell forever!

MARCELLA.   Farewell, oh best of cities!

FABIUS.   Farewell, crown of the world!

FABRITIUS.   Farewell, mother of letters!

SERGESTUS.   Farewell, generous homeland!

SALUSTIUS.   Farewell, light of captains!

ALBINUS.   Farewell, temple of the gods!

CELIUS.   Farewell, image of heaven!

OCTAVIUS.   Beloved Rome, I don't blame Caesar nor those who bear his maces, who have indeed been equal to him in mercy. I blame Genesius, who chose to perform the end of his life just at the moment of our richest opportunity. Thank God, the tragedy of his death stopped at the best character without touching

the rest of us. Comrades, how shall we put on plays successfully when we're bereft of the supreme actor? Who can play
Adonis as he did, with such grace and skill, such sprightliness
and wit?

MARCELLA.   You're the only one in the world, Octavius.

OCTAVIUS.   And tell me, who will play Paris at the destruction of
Troy?

FABRITIUS.   Fabius can; he's good at learning new parts.

MARCELLA.   We'll just have to learn some new plays, but in the
meantime, we can still go on performing the old ones.

OCTAVIUS.   Slow down, friends, for we've come to the Field of
Mars, which is the theater where Genesius is performing his
life and death this afternoon.

MARCELLA.   He's saying his speech at the end of the last act to the
people.

[A curtain is drawn, revealing Genesius impaled]

GENESIUS.   People of Rome, hear me: In the world I played its
wretched stories, its vice and wickedness throughout my life. I
was a pagan and worshiped pagan gods. Then God received
me. I'm now a Christian actor. The human comedy, all meaningless nonsense, is done. I've played the divine comedy
instead. Now I go to heaven to receive the reward for my
exceptional faith, hope, and charity. I owe them to heaven, and
heaven owes me for them. Tomorrow I'll see the sequel to this
play.

OCTAVIUS.   Here ends the play of the supreme actor.

The End